Neptune's Laboratory

Neptune's Laboratory

Fantasy, Fear, and Science at Sea

Antony Adler

Harvard University Press

Cambridge, Massachusetts, and London, England | 2019

Copyright © 2019 by the President and Fellows of Harvard College
All rights reserved
Printed in the United States of America

First printing

Library of Congress Cataloging-in-Publication Data

Names: Adler, Antony, author.
Title: Neptune's laboratory : fantasy, fear, and science at sea / Antony Adler.
Description: Cambridge, Massachusetts : Harvard University Press, 2019. |
Includes bibliographical references and index.
Identifiers: LCCN 2019009792 | ISBN 9780674972018 (alk. paper)
Subjects: LCSH: Oceanography—History. | Oceanography—International
cooperation—History. | Marine resources conservation—History. |
Ocean—Public opinion. | Ocean and civilization.
Classification: LCC GC57 .A325 2019 | DDC 551.46—dc23
LC record available at https://lccn.loc.gov/2019009792

For Judy and Dopher

Contents

	Introduction	1
1	Discovering Wonder in the Deep	13
2	Marine Science for the Nation or for the World?	46
3	Scientific Internationalism in a Pacific World	74
4	Cold War Science on the Seafloor	101
5	Ocean Science and Governance in the Anthropocene	135
	Conclusion	166

Notes *175*
Acknowledgments *231*
Index *233*

Neptune's Laboratory

Introduction

> The sea presents unfailing images of the *possible*—this, for me, is what sums up all its enchantment.
>
> Paul Valéry, Mer, Marines, Marins

> An oceanographic expedition is the crystallization of an idea; it begins and ends in the mind with a little sailing around in between.
>
> Henry W. Menard, Anatomy of an Expedition

In a scene near the end of *Moby Dick* (1851), Captain Ahab peers over the side of the *Pequod*, arrested by the sight of his own elusively shifting shadow on the water. In a moment of self-reflection he contemplates the "close-coiled woe" of his life, the privations and perils suffered during forty years at sea, and feels as if he were "Adam, staggering beneath the piled centuries." He drops a single tear, of which Melville remarks, "Nor did all the Pacific contain such wealth as that one wee drop." Ahab's shadow sinks beneath his gaze, "the more that he strove to pierce the profundity" of the "unsounded sea," and he turns from looking into the sea to look into the human eye of his first mate. Rockwell Kent illustrated the conclusion of the chapter with an image of a human face reflected by moving ripples of surface water. Although Melville was not writing about science or oceanography, the scene can serve as an allegory for interpreting the history of the marine sciences. For nearly two centuries, humans have studied, probed, and described the oceans, seeking to "pierce the profundity" of their depths. In exploring the seas, they have learned about the ocean environment, to be sure, but the history of the marine sciences also reveals something about scientists and humanity itself. In striving to piece together the history of marine science, we, like Ahab, may shift our gaze to look into human eyes. In this book, I explore how scientists, politicians, and publics have invoked the marine sciences and ocean environments in discourse about the fate of humanity and of the planet—debates that have conjured fantasies about utopian futures, as well as fears about human frailty and humanity's ultimate demise.

Modern conceptions of the future date only from the early nineteenth century. As the historian of science Iwan Rhys Morus has argued, in the early 1800s the Industrial Revolution led to "new attitudes towards progress, shaped by the relationship between technology and society." As a result, "people started thinking about the future as a different place, or an undiscovered country."[1] At the same time, the past was increasingly understood as "bygone and lost."[2] Ideas about what exactly an imagined future would be have been shaped by shifting social, political, environmental, and scientific concerns. Even as the global population surpassed its first billion, human impact on the environment, and doubt about the sustainability of human life itself, became a focus of anxious scholarly debate.[3] The opacity and mirroring qualities of the ocean, as well as its abundance of obscure life, make it an ideal screen for human projections of fear and hope. It thus comes as no surprise that, from the inception of modern oceanography, naturalists forging this new, "modern" discipline often framed their efforts as bringing forth desired outcomes for the human future or staving off impending cataclysms. In interpreting the history of the marine sciences, I follow the advice of the historian Denise Phillips, that "rather than assuming we know what science is, and searching through history to find places where people are practicing it, we will need to pay attention to how people themselves described the cause that captured their allegiance."[4] Hence, the theories and methods scientists use to gather new information and describe the features of oceans and their biota are important, but so are the publicly communicated rationales and imagery undergirding such scientific work. What concerns and causes have captured the allegiance of marine scientists? How have they sought to secure public engagement and support? What can we learn about the history of science at sea by attending to the imaginations that have motivated and shaped it?

Moreover, as historians have gradually come to appreciate in their studies of other disciplinary branches of science, a history of oceanography is incomplete if we focus only on the scientists. The ocean has long inspired painters, writers, and poets. But politicians, too, and publics invested in marine science have turned to it as a vast canvas on which to paint their fantasies and fears for the future. This book sets out to address issues such as these: What imagined futures have been embedded in marine science from its inception in the nineteenth century to the present? What anticipations and anxieties have been intertwined with particular oceanographic projects—expeditions, institutional developments, public outreach?

Histories of Oceanography: Seeking the "Cat behind the Grin"

The history of oceanography is a small but now well-established area of scholarship. Many of its founders, working in the 1960s, were oceanographers themselves or, in several cases, had close familial connections to oceanographic work. The first International Congress on the History of Oceanography took place in 1966 in Monaco, long an institutional center for marine science, and the president of that meeting was a French naval officer and marine scientist, Jules Rouch (1884–1973).[5] I point out the close craft and kinship ties between founding scholars and their subject of study not to discount their pioneering contributions but to explain why the history of oceanography has been slow to adopt some of the methodological approaches that have reshaped other branches of history since the introduction of social and cultural analysis in the 1970s. Histories of oceanography have long retained an emphasis on periodization, first discovery, and "great men" on the wane in other branches of the history of science.

Since the late 1990s, scholars seeking to move the history of oceanography beyond a teleological and progressive narrative have continuously struggled to define the limits of their area of study. Historians studying marine science in the twentieth century tend to distinguish between physical oceanography and marine biology, even if these categories do not always neatly correspond to the actual division of labor by marine scientists. As Christiane Groeben observes, "Oceanography is not a straightforward academic discipline. Rather, it is a complementary interaction of places, be they ships or labs, of people with diversified missions, and of instruments that mediate between the ocean and its observers and patrons."[6]

One of the reasons oceanography resists delimiting definition is that it operates on an enormous geographic scale. It is a science dedicated to studying three-fourths of our planet. As such, it has been one of the most interdisciplinary of efforts. Marine scientists have integrated geology, chemistry, physics, biology, and engineering in their investigations. And as a result, like its object, the history of oceanography too evades simple definition. "Trying to define the history of oceanography is akin to finding the cat behind the grin," wrote the preeminent historian of oceanography Eric Mills; "the closer one looks, the more one sees the grin and the less the cat." For these reasons, I use the terms "marine sciences" and "oceanography" to denote a broad range of practices deployed to gain a better understanding of the aquatic environment. These terms can be used to describe biological and physical oceanography but also fisheries science and underwater engineering. In taking this approach I follow the guidance of Eric Mills who, in 1990, wrote,

> Oceanography, as a discernable science and professionalized field, hardly existed before about 1910. Late in the nineteenth century, when the ingredients that gave rise to oceanography—marine biology and fisheries science, physics and chemistry, scientific expeditions and international science—were being stirred together, marine scientists thought of themselves professionally in a variety of ways, but rarely as oceanographers. As historians of oceanography we can either confine ourselves to the past few decades, when a definable field of oceanography existed, or make the more daring and significant choice of looking back to earlier, more puzzling times. Definitions turn out to be cramping, inhibitory, and worst of all, an unadventurous aid to our scholarship. History of oceanography is what historians of oceanography write about.[7]

With regard to the history of the marine sciences, historians have been slow to abandon what might be called the "great ship narrative."[8] This is undoubtedly in part due to reliance on official expedition narratives and scientific reports as primary source materials. But as Helen Rozwadowski has argued, the expedition or ship narrative was promoted by scientific entrepreneurs themselves to create a venerable tradition for their own enterprise.[9] The British *Challenger* expedition (1872–1876) is still described in some contemporary textbooks as a singular event that gave rise to the modern science of oceanography. And other great ships afforded prominence—to name but a few—include the British *Porcupine* and *Lightning*, the French *Talisman* and *Travailleur*, the German *Meteor* and *Valdivia*, and the American *Blake* and *Albatross*. In the many expedition accounts written by oceanographers on which historians have relied, the research vessel is given both prestige and agency; thus, historians have been led to foreground research vessels in their own accounts.[10] Attention to ships as scientific spaces is not to be avoided; indeed, it is critical.[11] But ships must be incorporated into a broader account of the many different spaces in which marine science has been conducted and discussed. Marine science was birthed not only in landlocked marine stations and universities but also in museums, fairgrounds, and homes. We miss an important part of the story if we focus solely on marine scientists while ignoring the larger world of interests supporting their work.

Oceanography requires data collection carried out in the field and focused on a specific, bounded geographic space, albeit an enormous one—the oceans (including their shores, surface, and depths).[12] It differs from many other field sciences by the degree to which data collection requires technological mediation; in this sense it can be compared to meteorology or space exploration. Only with the spatial turn in the early 1990s did scholars begin to develop new interpretive frameworks for accounts of scientific practices in the natural world. Thus, although the importance of laboratories as spaces for the production of scientific knowledge has long been acknowledged, it is only since the late 1990s that historians have given increased attention to the field as a crucial site of knowledge production. Still, most treatments of field sciences have dealt with terrestrial rather than marine topics.[13]

As with every scientific discipline, marine sciences have been decisively shaped by diverse economic, social, and political contexts. Although it is imperative to move beyond national histories of oceanography, historians gain important insights by recognizing national differences in forms of promoting and organizing marine science. The development of British marine science does not fully resemble that of French marine science, nor has French marine science mirrored its American counterpart. The case studies presented in this book focus on three countries: Britain, France, and the United States. Limited to these three national contexts, the goal is not to present a holistic transnational account but rather to show how disciplinary changes have arisen both within and across political boundaries. This has required an analytic approach that may be profitably adapted to investigate oceanographic developments in national contexts other than those treated here.[14]

Historians of marine science have benefited from a long tradition of maritime history that has persuasively shown why oceans must not be subordinated to terrestrial space as decisive arenas of human activity. Historians of marine science have also taken cues from trends in social, cultural, and environmental history—all of which have taken a watery turn.[15] Scholarship in the history of oceanography published in the first two decades of the twenty-first century demonstrates the interplay between developments in oceanography, changing state interests, forms of military support, and broader cultural shifts in Western society.[16]

Following what Elizabeth Deloughrey calls the "oceanic turn" in social sciences and humanities, historians are increasingly acknowledging the ocean as an environment worthy of historical attention—an environment, in the words of Helen Rozwadowski, "inextricably connected to, and influenced by, people" and "known through imagination as well as through direct experience."[17] Scholars of the history of fisheries have shown that the histories of science and the environment can be mutually revealing.[18] Stock depletions give added urgency to the study of past and present science, as well as uses, of fisheries resources.[19] Environmental historians describe a new body of scholarship as contributing to "an amphibious history," part of a broader effort to close "the blue hole in environmental history."[20] Yet even if, as W. Jeffrey Bolster has observed, "the past has never looked so watery," the ocean's third, vertical, dimension, and human engagement with it, remains understudied.[21] The history of human entanglement with the marine environment must be considered with attention to oceans as three-dimensional spaces, complete with creatures, currents, topographical features, and shorelines.[22] Oceans offer historians opportunities for transnational comparison and for melding analytic approaches of environmental history and the history of science. I take up this approach by scrutinizing the history of the marine sciences in relation to changing ideas about the import of the marine environment for the human future.

Practice and Place in the History of Oceanography

To understand the rise of oceanography as a modern science, we must link histories of changing instruments, ships, and places of work more closely to histories of the practices that invent, choose, deploy, and interpret them. Such an approach to science in practice turns our attention to the cascade of instruments and technologies that bring oceanic phenomena into view. But it also emphasizes the conditions of their use—the local circumstances, tacit knowledge, and improvisations that shape scientific work. Focus on practices also introduces a wider range of practitioners, amateurs as well as professionals, into our histories, bringing into view a fuller range of the activities that end by forging scientific knowledge about the ocean. It also encourages us to look at the motives, hopes, and fears undergirding this full range of work.

Historians of science in the field have, since the 1990s, given increased attention to large-scale field sciences first developed in the nineteenth century—oceanography, meteorology, and climatology.[23] Robert Kohler explains, "Perhaps the most noteworthy recent trend in the history of field science is the burgeoning interest in large-scale scientific practices, in sciences of space, science pursued on a regional or global scale. It's history of the 'long degree,' one might say . . . and it's becoming a major feature of the history of science."[24] Furthermore, as Kohler and others have shown, scientists have selected field spaces for their controllable, lab-like features.[25] This conflation of field and laboratory, taking place over the last two hundred years, has been a salient feature of oceanographic science.

Building on these new perspectives and insights, this book also draws on scholarly interest in histories of the future. The history of science fiction is now an established area of literary studies, but historians of science and technology are also giving attention to imagined futures of the past.[26] For example, Patrick McCray writes about the role of "visioneers" in the 1970s and 1980s to explain how utopian predictions and speculative technology led to the establishment of collaborative scientific communities and real-world technical developments.[27] Similarly, scholars in the interdisciplinary field of science, technology, and society have developed the theoretical concept of "sociotechnical imaginaries," defined by Sheila Jasanoff as "collectively held, institutionally stabilized, and publicly performed visions of desirable futures, animated by shared understandings of forms of social life and social order attainable through, and supportive of, advances in science and technology."[28]

In sum, the following account of the history of the marine sciences from the early nineteenth century to the present aims to move beyond teleological descriptions of first discovery and great-ship narratives focused on national contexts by borrowing analytic methods developed by environmental historians and historians of science. Prioritizing attention to imagined utopian and dystopian futures enhances our understanding of technological innovations, institutional developments, and international scientific collaborations that have shaped marine sciences. In surveying this history, we meet a diverse cast of characters: professional scientists and politicians but also mariners, amateur explorers, sportsmen, showmen, a prince, and tinkerers.

Chapter Overview

Chapter 1 presents an overview of marine science in the early nineteenth century. I explain how the oceans became the subject of popular and political discourse on both sides of the Atlantic. In the first half of the nineteenth century, technological developments rapidly transformed accessibility to the oceans—for scientists and members of the public alike. Not only could naturalists more easily go to sea; they might even do so in relative comfort—taking part in a practice increasingly seen as a pleasurable pastime by social elites.[29] Public interest in penetrating the mysteries of the ocean realm was boosted by the invention of home and public aquaria and the rapid industrialization of fisheries—which quite literally brought the ocean inland.

Chapter 2 concerns the development of European coastal marine stations in the late nineteenth century. Although the French government did not support large-scale oceanographic ventures, French marine stations garnered support as an aid to national fisheries and as the embodiment of a democratic vision of science: one connecting national glory with broad access to education. An examination of the French marine stations reveals the importance of institutional networks for the diffusion of scientific practice and their role in promoting some fields of study over others. The explosion of marine stations during this period helps explain the rise of coastal marine zoology in France and the comparative dearth of research in physical oceanography. A particular institutional context, and the scientific culture it fostered, had long-term ramifications for the development of marine sciences in France, where shallow-water research took precedence over state-sponsored deep-sea expeditions. Early proponents of the marine station movement appealed to international collaboration in an effort to widen the base of their political and financial support, but a growing nationalism threatened this project. In particular, rivalry between France and Prussia limited collaboration and stunted the growth of marine science in France.

Prince Albert I of Monaco, a towering figure of late nineteenth-century oceanography, was lionized by scientists for his contributions to the emerging science and, most importantly, for his achievements as its promoter and patron. Albert adhered to a total conceptualization of evolutionary life, in which the ocean was a source of vital powers ultimately leading to enlightened scientific governance of human societies and in which science was inextricably bound to ethical engagement. Albert's efforts to join oceanographic work with political internationalism and to forge transnational cooperative programs in marine science are illuminated by appreciation of his anomalous monarchical status. The institutions he founded, the Oceanographic Museum at Monaco and the Oceanographic Institute in Paris, bolstered French marine science in the absence of comparable French government support. They melded physical and biological research under a humanist program for marine research. Albert's social network as a regal head of state was crucial for the transnational promotion of oceanography. The dependence of this monarch's scientific program on dynastic lineage, however, also proved to be its greatest weakness.

Chapter 3 looks at aspirations for international collaboration in marine science. A key theme of late nineteenth-century rhetoric on the topic of oceanography concerned its promise as a focus of peaceful international cooperation. Whereas some programs for collaborative oceanic research were established in the period leading up to World War I— notably the International Council for the Exploration of the Seas— large-scale scientific internationalism in marine science made headway beginning only in the interwar period. Scientific internationalism of the interwar period established practices and gave voice to an ideology that later reemerged during the Cold War. Historians of oceanography have argued that the need for international cooperation in marine science derives from the geographic scale of the field under study.[30] The world's oceans encompass a territory extending beyond the claims of any nation, and the global scope of observations makes oceanography "an inherently international endeavor."[31] There is a sharp contrast between the state-sponsored, nationalist expeditions of the late eighteenth- and early nineteenth-century voyages of exploration and the collaborative oceanographic ventures that were succeeding them by the late twentieth century.

Scientific internationalism was central to the imagination of a unified Pacific World in the early twentieth century. In the aftermath of World War I, the Pacific Ocean promised a new terrain for peaceful international collaboration. The rhetorical appeal of cooperative transnational science, in conjunction with the internationalist movement during the interwar period, played a crucial role in shaping how the Pacific was imagined and described by commercial interests of the region as well as by scientists seeking public support for systematic, large-scale research in the Pacific basin. This vision was given further expression in the design and discourse surrounding mid-twentieth-century fairs, specifically the 1939 Golden Gate International Exposition in San Francisco and the 1962 Seattle World's Fair.

The dream of an internationalist program in the marine sciences suffered a major blow with the start of World War II. Yet even while the war helped institutionalize military patronage for science in the Pacific, internationalism remained an important component of science carried out on ocean-wide scales. Subsequently, during the Cold War, internationalist rhetoric was once again an integral component of popular science presentations meant to bolster public support for big science.

Chapter 4 examines these latter developments, taking as a case study a multiyear scientific research program named Project Sea Use. A collaborative effort of private industry, local government, and the US Navy to map and install instruments on the summit of a seamount off the coast of Washington State, Project Sea Use reveals how Cold War marine science built on government efforts to establish territorial jurisdiction on the continental shelf. As in outer space, the Cold War stimulated competition for supremacy in the oceans, often referred to at this time as inner space, and fantasies that human survival might depend on colonization of the aquatic frontier. The invention and marketing of scuba in the 1950s—a technology that opened the marine frontier to scientists and amateurs alike—fueled this vision and even birthed dreams of modifying the human species for survival in the aquatic realm.

Much has been written about the role of the US Navy as a patron of oceanographic science in this period, specifically through collaborations with Scripps Institution of Oceanography in California and Woods Hole Oceanographic Institution in Massachusetts. But similar collaborations with institutions in the Pacific Northwest have been overlooked. A Cold War history of oceanography that incorporates attention to the Pacific Northwest adds to our understanding of the history of American marine science, demonstrating how national policy—and fears of losing out in a race to the seafloor—played out in local arenas.

Chapter 5 discusses how boundaries between field and laboratory become blurred in contemporary marine science. Oceanographers now use fiber-optic cables, remotely operated vehicles, satellites, and robots to observe natural phenomena in environments where humans could not survive without the help of technology. While technological innovations have rendered the boundary between field and lab increasingly porous, oceanographers of the present share with their predecessors, the naturalists of the early nineteenth century, a desire to transform the marine environment into a more easily accessible, legitimate space for scientific experimentation and analysis. Scientific and technological innovations in modern times have informed our understanding of human-caused threats to the long-term health of the marine environment and shaped contemporary imaginations of the human, as well as of the oceans', future.

In her chronicle of the Scripps Institution of Oceanography, Elizabeth Noble Shor begins with the straightforward statement "Oceanography is not so much a science as a state of mind."[32] This is a key insight. If we accept the premise, which historians of science have long asserted, that scientific work, like all *human* activity, is shaped by human emotion and imagination, we will be alert to the importance of dreams, fantasies, and fears in its development and outcomes. In addition to offering an account of theoretical advances, technological innovations, and empirical discoveries, the history of marine science should include attention to the human emotions and efforts of self-understanding that have accompanied the struggle to "pierce the profundity" of Neptune's laboratory.

1

Discovering Wonder in the Deep

> The sea ... affords an almost endless variety of subjects for pleasing and profitable contemplation, and there has remained in the human mind a longing to learn more of its wonders.
> Matthew Fontaine Maury, The Physical Geography of the Sea

> You have carried out your work as far as terrestrial science permitted you. But you do not know all—you have not seen all. Let me tell you then, Professor, that you will not regret the time passed on board my vessel. You are going to visit the land of marvels.
> Jules Verne, Twenty Thousand Leagues under the Sea

In the second half of the nineteenth century, Europeans and North Americans gained a newfound appreciation for the open ocean as a vast three-dimensional space worthy of scientific exploration—as a "destination rather than a byway or barrier."[1] This scientific appreciation of the oceans occurred because, from the inception of modern oceanography, a literate middle- and upper-class nineteenth-century public—itself a product of industrialization—became invested in ocean exploration. Much like Victorian meteorology, nineteenth-century marine science captured the public imagination and (to a limited extent) invited public participation. As with meteorology, the scale of the work of gathering and assembling data about the vast marine environment was beyond the financial or logistical abilities of any single individual or even nation. This limitation fueled the creation of transnational scientific collaborative networks and led to publicized debates about national priorities, public funding for research, and scientists' first loyalties.[2] Naturalists drawn to marine science cultivated public interest in their work as a tactic for encouraging government support for their costly deepwater expeditions.

This chapter examines the growth of marine sciences in the second half of the nineteenth century: from coastal dredging practiced by amateur naturalists to state-sponsored deepwater oceanographic expeditions and to international fisheries exhibitions. It charts the paths by which marine science became the subject of popular, political, national, and international practices and discourse on both sides of the Atlantic.

As we will see, the actions of scientists alone cannot satisfactorily explain the multifaceted nineteenth-century cultivation of knowledge about the sea. Oceans were brought to the forefront of popular imagination and commercial and political attention by a convergence of factors. On both sides of the Atlantic people were increasing their consumption of fish, spending time at the seashore for health and recreation, reading maritime literature, and cultivating new forms of leisure activity in amateur science, yachting, and aquaria. Crucially, they also began to debate the environmental and political future of ocean environments—primarily with respect to the future of national competition over deepwater fisheries. To explain these developments I first examine why nineteenth-century scientists turned their attention to marine spaces and then explore how nineteenth-century publics encountered the ocean without ever going to sea: in private and public aquaria, in discourses about national naval power and commercial food supplies, and finally, at the fairground.

When Scientists Shunned the Sea

By 1800 most of the earth's navigable coastlines had been charted, and vast portions of its land mass had been "brought into the web of global commerce."[3] In the following decades, although scientific attention increasingly turned to the sea, marine science remained in its infancy. Naturalists whose work touched on conditions in the ocean were still more likely to rely on the testimony of others than to go to sea themselves. Prior to the voyages of discovery in the late eighteenth century, naturalists showed little interest in studying the physical properties or biology of marine environments. Writing in 1671, for example, Robert Boyle, the English natural philosopher celebrated for his air-pump experiments, wrote in a treatise on changes in pressure underwater, "I do not pretend to have visited the bottom of the sea, ... and [it is a] great rarity in these cold parts of Europe to meet with any men at all that have had at once the boldness, the occasion, the opportunity and the skill to penetrate into those concealed and dangerous recesses of nature."[4] Boyle had nevertheless sought out the testimony of "divers," "pilots," and "other navigators."[5] As early as 1663, however, Boyle's assistant Robert Hooke had begun to design instruments for improved sounding and seawater collection. Instructions for the construction of these instruments were later included in an appendix to *Directions for Seamen Bound for Far Voyages,* published in 1666 in the *Philosophical Transactions* of the Royal Society.[6]

Thus, for the most part, naturalists continued to acquire their information about the oceans through secondhand testimony. In a paper on the trade winds, British astronomer Edmond Halley wrote in 1687, "If the information I have received be not in all parts accurate, it has not been for want of inquiry from those I conceived best able to instruct me; and I shall take it for a very great kindness if any master of a ship, or other person, well informed of the nature of the winds ... shall please to communicate their observations thereupon."[7] Halley, unlike Boyle, did spend some limited time at sea, in 1698 commandeering the naval vessel *Paramour* for a scientific expedition to study magnetic variations in the South Atlantic. The venture was nearly aborted when the ship's officers threatened to mutiny, complaining of Halley's lack of qualification as a mariner.[8] Halley persisted, however, in attempts to venture even farther than seamen into the aquatic realm. He experimented with the use of a diving bell for salvage work in the Thames and off the Sussex coast, although not with the aim of advancing natural philosophy.[9]

Practical and utilitarian ventures could nevertheless lead to philosophical inquiry, however. Italian military officer Luigi-Ferdinando Marsigli (1658–1730) was unusual among naturalists of the late seventeenth century for wanting to carry out his experiments firsthand. His interest in hydrography, and his insistence on conducting his own investigations, can be understood by considering his military position and training. Sent on a diplomatic mission on behalf of the Venetians to Constantinople in 1679, Marsigli learned from local fishermen of the existence in the

Bosporus Strait of a deep current flowing in reverse to the surface current. Unsatisfied with secondhand testimony, he set out to observe the phenomenon himself, using white-painted corks that were visible when submerged on a weighted line.[10] He followed up these experiments by measuring current speed, water density, and changes in the tides.

As his biographer John Stoye argues, Marsigli's interest in "hydraulic matters" was in keeping with long-standing preoccupations of his Italian countrymen: control of rivers, engineering of irrigation channels, and management of flooding. The Venetians, in particular, had long sought means of controlling the accumulation of silt in their lagoon. To these engineering problems, Marsigli added his awareness of the military applications of hydraulic study.[11] Although his interests encompassed many aspects of natural history (he was particularly fascinated by coral), his work in hydrology encompassed both marine and freshwater bodies of water. The Great Turkish War brought Marsigli to Hungary and the banks of the Danube. Here he was charged with managing the movement of Habsburg military forces back and forth across the river—a problem that involved temporary bridges, naval battles on the river, and detailed surveys of depths and current speed, using instruments he had developed in his study of the Bosporus. As Stoye explains, "He watched the Danubian waters affecting the conduct of a campaign at almost every stage."[12]

As a military officer, Marsigli viewed aquatic spaces as a stage for adversarial contest. In his view, given what was at stake, empirical study rather than philosophical conjecture was essential. Traveling widely in the course of his military career and fluent in Latin, Italian, and French, Marsigli cultivated contacts with naturalists throughout Europe. After rising to the rank of general, he was demoted in 1704 and went into exile in southern France, where he carried out his most important marine work—depth soundings in the Gulf de Gascoigne conducted between 1706 and 1707. The opus for which he is now most remembered, his *Histoire Physique de la Mer*, was published in Amsterdam in 1725. It is recognized as the first oceanographic tome, although the term "oceanography" would not gain currency for another one and a half centuries. In this work Marsigli praised Boyle as a pioneer marine naturalist but rebuked him for reliance on secondhand testimony:

I was obliged to search, for myself, something more solid than the dissertation of Robert Boyle, and not to arrest myself at all that mariners would have me believe. I had thus to think up a number of observations which, when combined, could compensate for the impossibility of using eyes and hands to gain knowledge of that truth one searches for beneath the waters.[13]

Marsigli repeatedly gathered information himself using instruments of his own design.[14] The final aim of his research into the seas, this Italian virtuoso explained to readers, was to discover and reveal the "anatomy of the entire globe."[15]

Almost a century later, most scientists were still relying on the testimony of others to gain information about the oceans. On the other side of the Atlantic, for example, Benjamin Franklin relied on mariners in order to produce his three (1768, 1782, 1786) charts of the Gulf Stream.[16] But norms of land-based scientific field practice were starting to change, and it was increasingly expected that naturalists should gather their own data. A burgeoning age of global oceanic voyaging and scientific exploration, beginning at the end of the eighteenth century, was to lead naturalists to travel farther afield and to spend time on vessels at sea—if only as a means for getting from one land-based field site to another.[17]

Initially entering the maritime world as passengers, pioneer seagoing naturalists still regarded the oceans as terrain best understood by navigators with the accumulated experience of a centuries-old craft. Thus, the deference to mariners' expertise voiced by Boyle and Halley is still echoed by Johann Reinhold Forster (1729–1798), the naturalist on Captain Cook's second voyage around the world. Writing in 1778, Forster apologized to his readers for not having recorded more scientific information about the oceans while traveling on Cook's ship, offering as his excuse the impropriety of intruding on another profession's expertise:

I might here have subjoined many other particulars relative to the ocean; I might have given some account of its currents, and of the different constitution of its bottom, where we had any soundings; of various tides; and of the dipping and variation of the magnetical needle; but I forbear to speak on these subjects, as they are partly the objects of the nautical observations, made by the officers.... To make accurate observations ... would therefore be very improper, to attempt a business so ably discharged by others; who with proper instruments for that purpose, more leisure, and with more command of assistance, had better opportunities of making more perfect observations on these subjects.[18]

Forster had good reasons to "forbear" interfering in the work of the officers. His position as expedition naturalist was due to the ouster of a prior candidate. When Sir Joseph Banks, who had accompanied Cook on his first voyage, demanded that the vessel for the second voyage be modified in ways calculated to serve his scientific research, Cook refused to embark, contending that the ship was no longer seaworthy. With the backing of the Admiralty, Cook prevailed. The ship was refitted to his specifications, and Banks was forced to withdraw from the expedition. Forster replaced him.[19]

Even as they began to venture into oceanic space, early nineteenth-century naturalists lamented the discomforts of life at sea and struggled to secure an accepted place for themselves in the traditional social world of the ship. "I hate every wave of the ocean, with a fervor, which you, who have only seen the green waters of the shore, can never understand," wrote Charles Darwin to his cousin while aboard HMS *Beagle* in 1837.[20] Darwin became fascinated by the work of the *Beagle*'s hydrographers and soon adopted their sounding methods for his own field analysis.[21] Yet overlap between realms of expertise, and differences in priorities, easily led to conflict between naturalists and the mariners on whom they depended, the naturalists often finding themselves in a decidedly hostile environment.[22] "It is a curious fact," wrote the British naturalist Thomas Henry Huxley in 1854, "that if you want a boat for dredging, ten chances to one they are ... otherwise disposed of; if you leave your towing-net trailing astern in search of new creatures, ... it is, in all probability, found to have a wonderful effect in stopping the ship's way, and is hauled in as soon as your back is turned; or a careful dissection waiting to be drawn may find its way overboard as a 'mess.'" Naval officers resented the imposition of outsiders demanding space and accommodation for activities they regarded as lying outside the purview of military or navigational missions and as lacking the honor associated with naval service.

> The men of easy routine—harbour heroes—the officers of "regular" men-of-war, as they delight to be called, pretend to think surveying a kind of shirking—in sea-phrase, "sloping." It is to be regretted that the officers of the surveying vessels themselves are too often imbued with the same spirit; and though, for shame's sake, they can but stand up for hydrography, they are too apt to think an alliance with other branches of science as beneath the dignity of their divinity—the "Service."[23]

"Science is not the service," Huxley lamented.[24] As a result of resistance to accommodating the work of naturalists aboard, "adventures ashore were mere oases, separated by whole deserts of the most wearisome *ennui*."[25] Working at sea was undeniably uncomfortable, socially isolated, and certainly dangerous. But despite these disincentives, as areas for land-based field collection became ever more well-traveled, naturalists turned to the seas as spaces where striking discoveries—and in consequence careers—might yet be made. Marine spaces once imagined as "deserts" came to be reconceptualized as veritable cornucopias of life—a promising new world of marvel and wonder. Charles Wyville Thomson, who went on to lead the scientific company of the British *Challenger* expedition, wrote in 1873, "I had long . . . had a profound conviction that the land of promise for the naturalist, the only remaining region where there were endless novelties of extraordinary interest ready to the hand which had the means of gathering them, was the bottom of the sea."[26] Soon, important technological developments were to reshape ocean travel and exploration in ways that placed such novelties ready to hand.

Technology Transforms the Seas

In the nineteenth century, technological advances like steamships, submarine telegraphy, and improved navigational instruments transformed oceanic space. In the first half of the century, meticulous current and wind charts helped decrease the sailing time of Atlantic crossings. However, it was only upon the arrival of the first commercially viable transatlantic steamship, *Sirius,* in New York in April of 1838, that newspapers declared the successful "annihilation of space and time."[27] When soon thereafter some early passenger steamships exploded, at great loss of life, journalists blamed these accidents on "a public mind" that had become "completely infatuated with a wish to be borne in the twinkling of an eye" from place to place.[28] Indeed, the arrival of steamships transformed the experience of ocean travel. A crossing of the Atlantic, which in a sailing vessel might take six weeks traveling east to west, could now be accomplished in less than two. The new technology opened ocean travel to mass population migrations, and it was via steamship that millions of Europeans reached the shores of North America in the course of the nineteenth century.[29] Initially, conditions of mass migrant ocean travel were miserable—in 1846 more than 20 percent of the passengers sailing from Ireland to the United States died in transit. In the following decades, however, Britain and the United States passed legislation setting basic standards for shipboard accommodations, and conditions improved.[30]

By midcentury some passenger steamers were built as pleasure craft. In 1868, for instance, a British newspaper ran an ad for "a pleasure cruise round the Mediterranean, in a first-class steamship, calling at several places of classical and scriptural interest."[31] Among a growing affluent class, yachting and sailing races became a favored pastime of the wealthy. Yacht clubs emerged in coastal towns of England, Ireland, Scotland, and New York. Millionaires competed with one another to acquire ever-larger and ever-faster steam yachts.[32] As experience at sea, as passengers or yachtsmen, became more common among the middle and leisured classes, ships also became more accessible and alluring spaces for scientific pursuits.

Like the steamship, deep-sea telegraphy captured the public imagination; submarine telegraph cables were "the grand Victorian technology."[33] As cables enmeshed the world, helping consolidate the powers of European empires over colonial territories, some contemporary observers cast the new communication technology in biblical terms, as a means of ushering in a new era of human unification and peace. "Men talk to-day o'er the waste of the ultimate slime," wrote Rudyard Kipling. "And a new Word runs between, whispering, 'Let us be one!'"[34] But submarine telegraphy had the added effect of leading the public to contemplate a hidden and until then largely unimagined terrain.[35] A writer for the *Times* of London enthused, upon completion of the laying of a transatlantic cable in 1858, "Over what jagged mountain ranges is that slender thread folded; in what deep oceanic valleys does it rest, when the flash which carries the thought of man from one continent to another darts along the wire; through what strange and unknown regions, among things how uncouth and wild, must it thread its way!"[36] The necessity of gathering ocean sounding data for the deployment of submarine cables directed scientific attention to the abyss itself.

The Dredge and the Moving Scientific Frontier

As terrestrial field sciences developed and land-based field collection became ever more intensively practiced and competitive, the seas beckoned as spaces where important biological discoveries might yet be made. The publication of Charles Darwin's *On the Origin of Species* in 1859 provided further stimulus. Naturalists looked to the oceans as potential habitats of Lazarus taxons, or missing links, previously known only from the fossil record. Darwin himself noted the similarities between fossilized and living crustaceans, theorizing that maybe "the productions of the land" had changed "at a quicker rate than those of the sea."[37] This idea gained momentum when naturalists—in particular the Norwegian Michael Sars—discovered crinoids in the 1860s, a creature previously known only from the fossil record.[38] Scientific careers had long been built on the identification of new species and the gathering of significant collections. As unexplored territory shrank in the later nineteenth century, naturalists turned to the oceans as a new frontier for biological field collecting.[39]

Commercial fishermen were the first to invent and use dredges, mechanisms developed to harvest valuable bottom-dwelling mollusks, like clams and oysters. The first naturalists to adopt dredging for collecting marine organisms simply appropriated unmodified fishermen's dredges—or better yet, hired fishermen to dredge for them. In essence, the procedure involved dragging a weighted net along the seabed and scooping up whatever happened to be caught in its path. The technique produced a bounty of specimens and quickly became the favored field-collecting technique for professional and amateur naturalists alike. As one British author wrote in 1849, "Among the amusements of the seashore there is, perhaps, none so capable of yielding a varied pleasure to a person whose taste for Natural History is awakened, as dredging, where it can be carried on under favourable circumstances."[40] Although shallow coastal dredging was relatively easy, deepwater dredging remained a difficult and costly undertaking. Unlike coastal shallow-water dredging, often done with a rowboat, it required a much larger ship.[41] But scientists like Edward Forbes (1815–1854) promoted the technique, extending its use to ever-greater depths.

Born on the Isle of Man, Forbes spent his youth studying a wide range of subjects: art, medicine, botany, and geology.[42] He also traveled widely, visiting Ireland, France, Algeria, and Norway. Yet he remained above all fascinated by the ocean. As he later wrote, "The investigation . . . of the provinces of marine life, [has] as yet been but little pursued, and there is no finer field for discovery in natural history, than that presented by the bed of the ocean. . . . The difficulties which attend the inquiry add to the zest of the research."[43] Forbes's work is best appreciated as part of a larger movement reorienting European science to deep ocean work.[44] Taking up the challenge of extending access to the totality of the marine environment, Forbes set out to master the work of dredging and quickly became an expert. In 1851 Charles Darwin praised Forbes as knowing "more about dredging than all the other naturalists in Europe put together."[45] An outspoken promoter for the use of dredges for scientific work, Forbes persuaded the British Association for the Advancement of Science to form a "dredging committee" in 1839. This committee oversaw a coordinated effort to investigate marine life in the seas surrounding Great Britain. To aggregate data gathered through dredging, they produced what became known as "dredging papers"—blank forms on which naturalists could note the time and place where dredging was conducted and the organisms recovered. Forbes also organized classes in marine biology that included practical training in dredging techniques.[46]

The success of these first efforts encouraged the British Association for the Advancement of Science to fund further dredging expeditions. Thus, in 1841 Forbes was offered a berth on HMS *Beacon*, embarked on a survey expedition to the Aegean. He spent the next year and a half in the Mediterranean conducting the most extensive dredging surveys yet undertaken. Through this experience Forbes became interested in the global distribution of marine life. He realized that depth was an important factor in determining the distribution of species, and he developed a system for classifying the regions from which he recovered organisms. On the basis of his findings in the Aegean, he designated eight zones of depth in which marine life could be found. These were not static areas, he argued; rather, each was a "scene of incessant change" continually reshaped by fluctuations in marine organism populations and shifting geologic processes.[47]

Unfortunately, the Aegean report included a critical error. Forbes observed that fewer organisms were recovered at greater depths and, on the basis of this observation, hypothesized that a zone with "zero animal life" "probably" existed below three hundred fathoms. This he named the azoic zone. As historian Philip Rehbock has argued, "None of Forbes' generalizations was hazarded with so little fanfare and had such far-reaching influence."[48] The azoic theory dominated discussions of marine life distribution for the next decade.

Forbes had not failed to consider the available evidence. He had observed that marine plants could not grow at depths where light no longer penetrated. He also observed sedimentary deposits on land that lacked fossils, concluding that these layers "might have been formed in the very deep sea" where life was not prolific.[49] Unknowingly, Forbes had also been dredging in one of the most oligotrophic—poor in nutrients and low in biodiversity—regions of the Mediterranean. And perhaps most consequentially, he had been hindered by technological limitations of the dredge itself, which in its earliest forms could not reach below three hundred fathoms. The azoic theory seemed entirely warranted. How could life exist in darkness, extreme cold, and under great pressure? Thus, the theory was widely accepted until a single event threw it into question.

In 1860 a marine telegraph cable laid in the Mediterranean at a depth of 1200 fathoms broke. When the cable was recovered, naturalists were astonished to find it encrusted with marine life.[50] Forbes did not live to see this discovery because, succumbing to illness, he died at the age of thirty-nine in 1854. Yet over the course of his short life he helped change the way naturalists understood the distribution of life in the oceans, and most importantly, he inspired a new generation of naturalists to adopt dredging and take up the study of deepwater biology. The stage was set for an ever-expanding scientific investigation of the seas.

Meanwhile, on the other side of the Atlantic, the primary promoter of marine research was the US government. With marine commerce at stake, it created two agencies tasked with surveying and charting coastal waters: the US Coast Survey, established in 1807, followed by the Depot of Charts and Instruments—a branch of the US Navy—in 1830.[51] By 1840 both agencies had integrated measurements of currents, temperature, salinity, and bottom composition into routines of sounding and charting.[52] Much of the work of these organizations dealt with the study of the Gulf Stream, which swept past the Eastern Seaboard and was of vital importance for the nation's maritime commerce and fisheries. However, other research pushed farther afield—most notably in the Pacific under the auspices of the US Exploring Expedition (1838–1842), commanded by the naval lieutenant and head of the Depot of Charts and Instruments Charles Wilkes (1798–1877).[53] Here too, commercial interests preceded scientific exploration, because New England whalers and traders were already crisscrossing the Pacific basin.[54] Wilkes's resultant publication on hydrography was delayed until 1873. It included a global wind chart intended to aid circumnavigational voyages. The US Exploring Expedition, inspired by the great voyages of discovery of the late eighteenth century, also embodied the nineteenth-century Humboldtian objective of global observation and precision measurement. This objective, however, was to prove elusive without international collaboration.

By midcentury, a US naval officer and director of the Naval Observatory, Matthew Fontaine Maury (1806–1873), who had published a series of wind and current charts in 1847 and the influential *The Physical Geography of the Sea* in 1855, began work to organize the first international maritime conference.[55] Held in Brussels in 1853 and attended by representatives from several European navies, the stated goal of this meeting was to devise "a systematical and uniform plan of meteorological observation at sea." From the very beginning, this was envisioned as a global internationalist project. Maury wrote in his final report, "The navies of all the maritime nations should cooperate, and make these observation in such a manner and with such means and implements, that the system might be uniform, and the observations made on board one public ship be readily referred to and compared with the observations made on board all other public ships, in whatever part of the world."[56] Maury long dreamed of transforming "every ship that navigates the high seas" into a "floating observatory, a temple of science."[57] Maury was a Southerner who sided with the Confederacy, however, and his naval career did not survive the Civil War.[58]

Oceanic biological research in the United States was further institutionalized in 1871 through the establishment of the US Fish Commission and in 1886 when an act of Congress established the Hydrographic Office as a division of the Bureau of Navigation.[59] Together with the Coast Guard and US Maritime Commission, the Hydrographic Office supplied navigational charts and coordinated the collection of oceanographic and hydrographic data.[60] Thus, by the end of the nineteenth century, several branches of the US government fostered a broad range of marine research. State support for marine science continued to be dominated by military and commercial concerns, however, and state-sponsored projects showed little interest in questions of biology or, for that matter, even fisheries management.[61] Naturalists who engaged in open-ocean dredging and collecting continued to hope that the work, requiring international support, might serve as a catalyst for peaceful diplomacy. But support for deep-sea biology required much greater public interest and involvement.

People Travel to the Shore and Sea Life Moves Inland

During the nineteenth century, the seaside became a place to visit for health and leisure, yachting and fishing flourished as favored pastimes of the rich, and maritime fiction, as embodied in the works of Robert Louis Stevenson, Herman Melville, Jules Verne, and, by the end of the century, Joseph Conrad, came into its own. Physicians prescribed trips to the shore for the revitalizing effects of cold water and salt air.[62] Artists, writers, and amateur naturalists were inspired by news of the scientific discoveries being made in the oceans. Romantic rhetoric cast the ocean as sublime space where individuals could test and reclaim their humanity.[63] And a lay public took new interest in marine life, as discovered on the fashionable shoreline, in shallow coastal waters, in Victorian drawing rooms, or

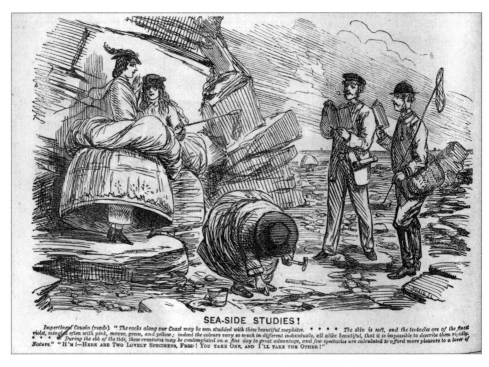

Figure 1.1 "Sea-side studies!" A cartoon from *Punch's Almanack* 38 (June 1860). Reproduced from a copy at St. Olaf College Libraries.

on their dinner plates (see Figure 1.1). As submarine telegraph cables were installed on the seafloor, popular fiction encouraged the reading public to imagine exploring the depths from their armchairs through such works of fiction as *20,000 Leagues under the Sea*. And as previously unknown species of fish were dredged from the sea, nineteenth-century middle-class families marveled at living ocean creatures in newly invented public and private saltwater aquaria. Interest in the marine environment was further encouraged by popular science writing, public lectures, and large exhibitions, as well as by a growing fishing industry working to expand consumer markets.

A new vogue for small household and large public aquaria made it possible to view living marine creatures indoors and inland, serving as miniature representations of marine ecosystems and spurring imagination of a diverse underwater panoply of life.[64] "The aquarium is an established household ornament," declared an 1856 English book on household interior design. "It graces the drawing-room, the conservatory and the greenhouse; is a welcome and highly-prized addition to the student's resources in the acquisition of knowledge; it extends the sphere of domestic education for the young, enlivens the solitary house of the invalid, and gives delight to everybody."[65] By midcentury, numerous publications covered all aspects of aquarium design and maintenance as home aquaria became an indicator of social status and wealth.[66] The craze for private saltwater aquaria began in Britain, spread abroad, and by the 1850s, associations of saltwater aquarists could be found in the United States and on the Continent.

Large public aquaria were also established at this time, the first in Regent's Park, London, in 1853. The Fish House was a popular attraction, although it was plagued by engineering problems, chief among them the apparent impossibility of controlling the temperature in tanks filled with salt water shipped from the coast. Urbanites of many countries were getting their first look at living marine creatures miles inland. Alexander Ussner, manager of the Vienna aquarium, recalled in 1860, "A cry of admiration and astonishment went literally not only around Europe, but around the whole civilized world when this first water menagerie was opened to the public."[67] Novel public aquaria, sometimes organized by private showmen, became popular on the other side of the Atlantic as well. A vivid description of the Boston Aquarial Gardens survives in the diary of eleven-year-old Sarah Gooll Putnam, who visited them in November 1861. She describes seeing fish, dolphins, and a whale in a tank (see Figure 1.2). The whale was "as long as two short men" and "as white almost as snow." Her diary entry is accompanied by a sketch showing a circular tank, complete with a fountain, surrounded by smaller tanks. On a second visit in February 1862, Sarah describes witnessing a performance orchestrated with the captive whale: "We saw the whale being driven by a girl. She was in a boat, and the whale was fastened to the boat by a pair of reins, and a collar, which was fastened round his neck. The men had to chase him before they could put on the collar."[68]

The American showman P. T. Barnum, responsible for the Boston aquarium, also built an aquarium at his New York museum in 1860. When the museum was destroyed in a fire in 1865, two captive belugas perished

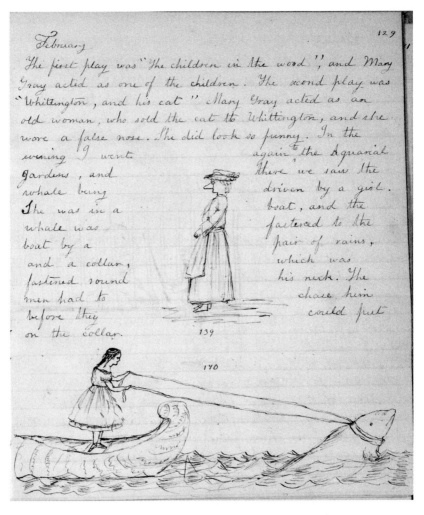

Figure 1.2 Illustration from Sarah Gooll Putnam diary (February 1862). Sarah Gooll Putnam diary, 129, Massachusetts Historical Society.

(two of nine belugas that died prematurely under Barnum's watch).[69] Barnum's partner in these ventures was Henry D. Butler, author of *The Family Aquarium* (1858), who described aquaria as a "necessary luxury in every well-appointed household, both of Europe and America."[70] Another one-time partner, the showman William Cameron Coup, opened the New York Aquarium on Broadway in 1876. The building contained a scientific library, a laboratory, and in the main hall, a tank six feet high and thirty feet wide intended for a whale. The first captured whale destined for the exhibit died before delivery, but Coup found a replacement.[71] At this time, even a dead whale could capture the public imagination—and generate revenue. In his memoir Barnum recounts that in 1855 he arranged for the exhibit of the refrigerated carcass of a twelve-foot-long "black" whale. This "created considerable excitement" and the profits allowed him to pay "the entire board and bill" of his family "for the season."[72] In the 1880s another decaying whale carcass, aptly named "the Prince of Whales," toured the United States by rail for over two years.[73]

Small home aquaria brought living marine creatures into a controlled environment. Thus, although appreciated for their entertainment value, they also provided instruments for the observations and experimentation of naturalists. As one aquarium book of instructions announced, "To the naturalist the aquarium opens up new studies of the choicest wonders of the deep sea," adding that it was to be hoped that "departments of zoology" that had "hitherto made the slowest and least satisfactory progress" would now experience a "sea-change" permitting the study of the "habits and economy even to the minutest particular" of sea creatures recovered thanks to the dredge.[74] Aquaria were enthusiastically compared to the telescope and the microscope, in opening "to our inquisitive gaze the hidden chambers of the deep."[75]

The individual most responsible for popularizing aquaria in Britain was a naturalist who specialized in collecting marine fauna. Philip Henry Gosse (1810–1888) was a prolific author of popular natural history; he was also deeply religious, advocating the study of science as a means of getting closer to God. In one of his memoirs, *A Naturalist's Guide to the Devonshire Coast,* he describes a visit to the seashore prescribed to improve his health. He passed the time by collecting specimens and delighted in studying creatures kept alive in makeshift tanks, describing the painted scallop, for instance, as "an object of unwonted beauty."[76] Later, reflecting on the study made possible by his experimentation with the new technology, he posed a rhetorical question to his readers: "Should these experiments be perfected, what would hinder our keeping collections of marine animals for observation and study, even in London and other inland cities? . . . I hope to see the lovely marine Algae too, that hitherto have been almost unknown except pressed between the leaves of a book, growing in their native health and beauty, and waving their delicate translucent fronds, on the tables of our drawing-rooms and on the shelves of our conservatories."[77]

Shortly thereafter, Gosse published a guide to aquarium maintenance, *The Aquarium: An Unveiling of the Wonders of the Deep Sea*. Included are guidelines for the types of marine creatures to be collected and housed together in order to maintain the health of all organisms in the tank. Gosse suggests, for example, that marine plants should be installed first, to oxygenate the water and that subsequently periwinkles should be collected to clean the glass.[78] He cautioned readers not to overcrowd their tanks, to "be moderate in [their] desire of dominion."[79] Aquaria maintenance required repeated visits to the sea, to collect organisms and fresh seawater. Gosse assured his readers, in London "sea-water may be easily obtained by giving a trifling fee to the master or steward of any of the steamers that ply beyond the mouth of the Thames."[80]

Gosse's work was instrumental in launching an aquarium craze throughout Europe. "After reading the book of Mr. Gosse," one writer remarked, "all the world wanted to possess an aquarium to verify his assertions and repeat his experiments."[81] Professional and amateur naturalists alike embraced the use of aquarium tanks for the study of marine zoology. Aquaria brought the ocean inland and indoors. And as nineteenth-century aquarium owners and visitors contemplated the sea life now made visible in private drawing rooms and public exhibitions, they could ponder the discoveries yet to be made in still-unfathomed depths. Probing that abyss, however, was to prove an enormously costly and technologically demanding venture.

HMS *Challenger*: "A New Experiment"

Charles Wyville Thomson (1830–1882) was born in Bonsyde, Linlithgow, Scotland, the son of a surgeon of the East India Company. He attended the University of Edinburgh to study medicine, but his interest in the natural sciences soon occupied all his time. In 1850, at the age of twenty, he was appointed lecturer in botany at the University of Aberdeen; three years later he became professor of natural history at Queen's College, Cork; in 1854 he took up a position as professor of geology at Queen's

College, Belfast, and soon also became professor of zoology and botany. In 1868 he moved to the Royal College of Science in Dublin. One of his students later recalled an inspiring lecturer "who had on his table a profusion of specimens of which he made incessant use, but spoke without notes."[82] Thomson's interest in the ocean first took form in Belfast. He went dredging with Edward Forbes in 1854 and visited the leading Norwegian marine biologist Michael Sars in 1866.[83] In 1868 he began his own dredging investigations, along with William B. Carpenter (1813–1885), a physiologist at the University of London.[84]

Thomson's turn to oceanography coincided with the breaking of a submerged telegraph cable between Sardinia and Algeria in 1860. The celebrated French naturalist Alphonse Milne-Edwards published a report on the resultant discovery of deep-sea marine life the following year, permanently putting to rest Edward Forbes's azoic hypothesis and announcing the deep sea as terrain for biological investigation.[85] With the support of the Royal Society, Thomson and Carpenter persuaded the Royal Navy to grant them use of the naval vessels HMSS *Lightning* and HMSS *Porcupine* to conduct dredging operations and take water temperature measurements off the coasts of Ireland, Scotland, the Shetlands, and Faeroes in the summers of 1868, 1869, and 1870. These initial efforts yielded valuable biological collections and put Thomson on the path that eventually led to his selection as chief scientist of the HMS *Challenger* expedition (1872–1876).[86] Thomson was elected to the Royal Society in 1869 and, in the following year, appointed to the Regius Chair of Natural History at Edinburgh—a post formerly held by Edward Forbes. Thomson and Carpenter now lobbied for state support for deep-sea exploration, and in 1871, with the support of the Royal Society and the British navy, established a Circumnavigating Dredging Committee. Four years later it was renamed the *Challenger* Expedition Committee.[87]

The voyage of HMS *Challenger*, commanded by the naval captain George Nares and undertaken at the prompting of Thomson, is often cited as the singular event that gave rise to modern oceanography.[88] A pioneering accommodation of shipboard science by modifications of vessel space changed the social relations of scientists and crew and paved the way for the development of the modern oceanographic research vessel as floating laboratory.[89] The events of the *Challenger* expedition have been covered in great detail elsewhere.[90] It suffices to observe here that although the expedition has frequently been said to mark the birth of oceanography, it can be argued that it is more accurately understood as a scale-up of established techniques.[91] Thomson and Carpenter had already completed successful deepwater dredging on the *Lightning* and *Porcupine*, and the British government was spurred to match and exceed the deep-sea work already carried out by the US Coast Survey. (British naturalists frequently cited the work of the Americans in efforts to shame their government into action.[92]) Thus, rather than interpreting the *Challenger* expedition as marking the "birth" of oceanography, it is better conceptualized as signaling the maturity of nineteenth-century oceanography as an established discipline and as "big" state-supported science.[93]

In reality, the project was not completely novel, but Thomson nevertheless described it as "altogether a new experiment."[94] Part of the reason for the *Challenger*'s primacy of place in the history of oceanography may stem from a rhetoric adopted by expedition members that inflated the grandeur of their achievement by minimizing the significance of prior marine investigations. "We had very indefinite and obscure ideas about the great deep sea. In our imagination it was the place of 'dark unfathomed cave,' the home of 'gem of purest ray,' or the abode of a mighty sea serpent," declared John Murray in his first public address on the results of the *Challenger* expedition, delivered in 1877. "In place of these imaginings we have now some definite knowledge."[95]

Even if the research techniques used were not as new as Thomson suggested, the scope of the work was certainly unprecedented. Outfitted with 144 miles of hemp rope for sounding and 12.5 miles of piano wire for recovering seabed samples, for a three-and-a-half-year circumnavigation of the globe, the *Challenger* probed the ocean's depths on a scale previously unimaginable. As a result, the expedition amassed an unprecedented collection of biological specimens.[96] Estimates are that four or five new species were discovered for each day the expedition spent at sea.[97] It is no wonder, then, that the *Challenger* expedition was instrumental in bringing the deep sea to the forefront of public attention, and not only in Britain. A journalist for the *New York Herald* hailed the expedition as "the most important that has probably ever left any country."[98] Even naturalists who had never gone to sea were eager to gain access to the expedition's prized specimen collections.

The British botanist Henry H. Higgins, writing in *Nature,* argued that duplicate collections ought to be sent to public museums in Great Britain and Ireland. He reasoned that "public tax-payers [had] willingly contributed towards the expenses of the late noble and successful expedition" and that putting the collections on public display would serve as "a recognition of their claim to share in the treasure trove." In advocating for the distribution of the collections, Higgins hoped further impulse "might be given to the study of natural science, and to the cordial support of plans for further expeditions of a like character." Such distribution, he argued, offered "a great national opportunity."[99] Duplicate collections for public exhibition were not, however, Thomson's priority. Having amassed an enormous number of specimens, many new to science, Thomson was keen to disperse them to experts for analysis. But here he faced a new problem. As Margaret Deacon has observed, "Whereas before the *Challenger* set sail relatively few people were interested in deep-sea work, now zoologists were eager to participate in what promised after all to be a rich and rewarding field."[100] Thomson decided that the work of cataloging the Echinoidea should be entrusted to Alexander Agassiz and Theodore Lyman, both of the Museum of Comparative Zoology at Harvard.[101] The sponges were to go to Oscar Schmidt and the Radiolaria and deep-sea medusae to Ernst Haeckel, both in Germany, and some of the Crustacea were to go to the Norwegian Georg Sars, son of Michael Sars.[102]

Alexander Agassiz conceived of his contribution to the collection's analysis as casting glory, by proxy, on the United States. In a letter posted from Edinburgh in December 1877 to the *American Journal of Science and Arts,* and reprinted in the British *Annals and Magazine of Natural History,* Agassiz happily announced that he was bringing over "the Echini, and perhaps some other group of Echinoderms; so that the United States will have their fair share of the work."[103] Not all were happy about this distribution of the *Challenger*'s collections, however. Agassiz's declaration raised the ire of Peter Martin Duncan, president of the British Geological Society. Writing to the editors of *Annals,* Duncan excoriated Thomson's distribution of the collections as "unjust and unpatriotic." British naturalists, he argued, had been "passed over with contemptuous neglect." From Duncan's perspective, that parts of the collections had been distributed to foreign naturalists was symptomatic of problems despoiling British science:

> For a great nation to send out expensive expeditions and then to distribute the results for determination and description to foreign naturalists, however distinguished, without considering and employing its own naturalists, is rather characteristic of this age of depreciating criticism; but it is a proceeding which can only be tolerated upon a preposterous application of the idea of catholicity in science and the fact of the incompetence of national investigators.[104]

Taking aim at Thomson, Duncan went so far as to argue that his "apparent ignorance of the work of his fellow countrymen would seem to disqualify him for his position."[105] The drama continued to play out in the pages of *Nature*, which published subsequent exchanges between Duncan and Thomson. Claiming to write on behalf of "many men" of science, Duncan explained that although they might "recognize the merits of those foreign gentlemen to whom you have sent collections . . . [,] we do not think that you are justified in giving them the results of the greatest natural history expedition which has ever sailed from this country." Ultimately, Thomson prevailed, thanks to the support of powerful allies—Joseph Dalton Hooker, Thomas Henry Huxley, and Charles Darwin—who defended the international distribution in a joint letter to the editors of *Nature*: "If this country can be shown to enjoy the unique distinction of possessing in every department of zoological research men at least as good as can be met with elsewhere, the advocates of a national science may find an argument in favour of having the work absolutely confined to Englishmen; but if we cannot assume a position which no other nation in the world would think of claiming, it is plainly in the interests of science that we should supplement from abroad those departments of research in which foreign workers may excel us."[106]

Despite this victory, internationalist sentiment in marine science was far from unanimous and slow to catch on.[107] The American navy captain George Belknap promised the naturalist William Healey Dall that American work in the Pacific would put "the 'Challenger' quite in the shade."[108] And when the British naturalists John Gwyn Jeffreys and Alfred Merle Norman were invited to join a two-week cruise on the French *Travailleur* oceanographic expedition in 1880, they were not permitted to work on the collections—a snub they claimed to have taken lightly, confident in the superiority of British marine zoology.[109] These tensions reveal the strains of a period of transition in oceanographic research in the latter half of the nineteenth century, when imperialist expeditionary models of mission and loyalty clashed with another conception of mission in which scientific expedience and an international scientific community were granted precedence over national loyalty and glory.

It took twenty years for all fifty volumes of the *Challenger* reports to be compiled and published. Thomson did not live to see the completion of the project because he died at the age of fifty-two in 1882. The herculean task of compiling the reports was completed in 1895 under the direction of fellow *Challenger* naturalist John Murray (1841–1914). Writing to Murray from Boston, in a letter that illustrates internationalist sentiment, Agassiz lamented the loss of Thomson: "All he has thought and written on [oceanography] cannot be found combined in any one man.... The death of my own brother could hardly be more heartily felt and I shall sadly miss his correspondence. He was one of the few scientific men [from] whom I liked to have frequent letters and to talk [about] our plans of the future and discuss the past."[110] The *Challenger* expedition and its resulting reports cost the British government a total of £171,000—making it one of the largest and most costly research projects undertaken by any nation prior to World War II.[111] These reports—the last of which was published in 1895—were distributed free of charge to scientific institutions around the world.[112] As Margaret Deacon has argued, "Thomson's achievement was to create a network of personal relationships which, in the absence of concrete recognition either academic or governmental, served as the first step towards building up an international community of oceanographers."[113] Certainly the international distribution of the collections for analysis, the ensuing correspondence, and the free distribution of the published reports—all orchestrated under the direction of John Murray after Thomson's death—indicate this shift. But while a fledging international community of naturalists was building around the oceans as a fruitful field for biological research, others were beginning to take an interest in oceanic research for purely commercial reasons. Scientists still knew little about the seas in the late nineteenth century, but the Industrial Revolution had already dramatically amplified commercial marine fisheries.

The Growth of Industrialized Fishing

Railroads, which carried nineteenth-century pleasure-seekers to the coasts, also transported living or freshly caught marine organisms inland. This revolution in transportation had two important effects: it boosted the commercial value of the marine fishing industry by opening new markets and driving demand for increased production, and it facilitated the work of inland naturalists who, previously hindered by geographic distance, could now study and experiment on freshly caught marine organisms. New markets for saltwater fish, permitted by expanding rail networks, stimulated the need to increase catch, and improved fishing vessels allowed fishermen to work in deeper waters. As fisheries expanded, and fleets of several nations harvested from the same regions of the North Sea and the Atlantic, fear of a coming depletion of fish stocks grew. In 1862, the MP for Sutherland, James Fenwick, calling for a royal commission to investigate the state of British fisheries, sounded the alarm: "The almost universal cry is that our fisheries are falling off year by year."[114] The commission was established in 1863. A solution to the threat of declining stocks was sought by some through the artificial manipulation of fish populations, but France, not Britain, was at the forefront of aquaculture research.[115]

British and French anxieties over the strength of national fisheries added to their rivalry because fisheries were understood as key to national power. Marine fisheries provided food to support an increasing population, and fishermen could be pressed into service aboard naval vessels—the critical military technology of the time. A satirical passage from *Punch* underscores contemporary British concerns over France's maritime military aspirations by describing "the ocean [as] a French lake and England at the bottom of it."[116] In France, one of the most important fishery expositions of the period was held in 1866 in Arcachon, a fishing port on the Atlantic seaboard. This exposition on fishing and aquaculture, organized by the local scientific society, led to the creation of a coastal biological station and marine museum-aquarium. In preparation for the exposition, the organizers published a lengthy questionnaire in the Bulletin de la Société Impériale Zoologique d'Acclimatation. This questionnaire was divided into three sections: "Natural History," "Technology," and "Aquaculture and Social Economics." Each section was headed by a thematic sentence: "dominate the fish of the sea"; "is aquaculture to the fishery what agriculture is to hunting?"; and "if agriculture furnishes us with soldiers, the fishery furnishes us with sailors."[117]

A British author provided his impressions of the exposition, voicing concerns about the weakened state of the British navy in the face of expanding French naval power. He suggested that the fisheries, now including aquaculture, were key elements of French military strategy:

> Aquiculture [sic] in every form is wrought so that it shall tend to "the glory and strength of France." Even the expositions at Arcachon and Boulogne are . . . visible tokens of the rôle which the Emperor would have populations . . . adopt. . . . The French Government is rapidly extending an industry which produces very large . . . revenues. This therefore, is not only strength and glory; it is internal wealth wherewith to support them.[118]

Expositions served to promote marine science in France, a trend that continued into the early twentieth century. Napoleon III was presented with the Arcachon exposition report in 1867 but had little time to implement its recommendations because he fell from power four years later, at the end of the Franco-Prussian War.[119]

The British too, keen to support their growing fishing industry, turned to expositions as a means of promoting technological innovation and encouraging fish consumption. Inspired in part by an international fisheries exposition held in Berlin in 1880, the Great International Fisheries Exposition opened in London on May 12, 1883. Remembered later as a "wretchedly cold and rainy morning," the opening ceremonies still drew fifteen thousand people. The great promenade where the ceremony took place was crowded with "well-dressed persons," and a dais, "festooned with herring-nets, and prettily adorned with tridents and banners," was set up for the royal family. Assembled in the great gallery devoted to the British fisheries were "four hundred representative fishermen from all parts of the kingdom, costumed in their jerseys and sou'westers, their overalls and sea-boots, precisely as they would be when daring the perils of the deep." The Prince of Wales delivered the opening speech:

> In view of the rapid increase of the population in civilized countries, and especially in these sea-girt kingdoms, a profound interest attaches to every industry which affects the supply of food; and in this respect the harvest of the sea is hardly less important than that of the land. I share your hope that the Exhibition now about to open may afford the means of enabling practical fishermen to acquaint themselves with the latest improvements which have been made in their craft in all parts of the world, so that without needless destruction or avoidable waste of any kind mankind may derive the fullest possible advantage from the bounty of the waters.[120]

In 1883, between May and October, 2,689,092 people visited the exhibition—a daily average of over 18,500.[121] Many nations sent materials for display; the US gallery alone assembled twenty-five thousand exhibits at a cost of $50,000.[122] The exhibition was so large that it was not uncommon for visitors to buy season tickets and make repeated visits. The exposition was designed to display all things even remotely related to the sea.[123] It included fishing apparatus, a picture gallery, relics recovered from shipwrecks, diving suits, stuffed sea birds, and the skeleton of a whale coated in phosphorescent paint. An enormous aquarium—the largest ever constructed—with thirty saltwater and nine freshwater tanks, was one of the most popular attractions.[124] In a small fish market visitors could purchase "a basket of fish, as a memento of their visit" (see Figure 1.3).[125] Or they could dine in the exposition saloon where, "barring only salmon, turbot, soles, and expensive shell-fish," almost any type of seafood could be obtained, including some exotic specimens. The exposition tapped Victorian fascination with the sea, but it also gave visibility to a growing urban market for seafood and to the rapid transformation of commer-

Figure 1.3 Fish market at the International Fisheries Exhibition. Frederick Whymper, *The Fisheries of the World: An Illustrated and Descriptive Record of the International Fisheries Exhibition, 1883* (London: Cassell, 1884), 105.

cial fisheries. As one American observer reported, "The press is full of plans for the practical outcome of the exhibition. Some of the editors expect to see fish cheaper; some, to see the cheaper kinds of fish coming into general use; some to see fish of all kinds more generally used; some, to see an immense increase in the yield of fisheries; some, to see legislation stricter and more strongly regulated."[126]

As fisheries grew, scientists and politicians made a habit of citing the health of the nation's fisheries and aquaculture as a metric for the vitality and power of the state. As with the development of aquaculture, France was also at the forefront of industrialized fishing, and it was there that the first modern steam trawlers were developed.[127] Early experiments with fishing vessels outfitted with steam engines were carried out in the late 1830s near the fishing port of Arcachon. By 1876 a French company, Société de Pêchérie de l'Ocean, was operating two commercial steam trawlers in the North Sea.[128] In France, the most important fisheries were for North Atlantic herring and cod, and the bulk of this catch was landed in the ports of La Manche, in Brittany, followed by ports of the Atlantic seaboard.[129] In descending order of economic importance, other large-scale French fisheries harvested sardines, herring, mackerel, mussels, anchovies, crustaceans, and oysters.

On the other side of the Atlantic, New England fishermen off the coast of Nova Scotia, seeing cod landings decrease by more than 50 percent between 1852 and 1859, placed the blame on large French fishing vessels that frequented the Grand Banks, each armed with longlines deploying thousands of baited hooks.[130] In an effort to stave off further declines to the fishery, the Americans established the US Fish Commission in 1871, under the direction of Spencer F. Baird, which advocated fish culturing. But technological advances continued to accelerate the depletion of wild stocks, and the arrival of the steam trawl in the 1880s was disastrous for the North Atlantic fishery.[131] The French scientist Julien Thoulet (1843–1936) observed when he traveled to the Grand Banks on a French naval vessel in 1890, "Since the fishery is carried out in the most harmful way possible . . . this source of wealth will soon be exhausted forever."[132]

In many parts of Europe, as on the east coast of North America, fishermen voiced their alarm about the decline of saltwater fish stocks. Particularly contentious was the use of trawls and driftnets.[133] In England petitions to the government resulted in the appointment of a series of royal commissions tasked with assessing the sustainability of British fisheries. The celebrated naturalist Thomas Henry Huxley was appointed to three of these commissions, starting in 1863.[134] The first report of 1866, written under Huxley's direction, contemptuously dismissed fishermen's testimony:

> Fishermen, as a class, are exceedingly unobservant of anything about fish which is not absolutely forced upon them by their daily avocations; and they are, consequently, not only prone to adopt every belief, however ill-founded, which seems to tell in their own favor, but they are disposed to depreciate the present in comparison with the past. Nor, in certain localities, do they lack the additional temptation to make the worst of the present, offered by the hope that strong statements may lead the state to interfere in their favor, with dangerous competitors.[135]

With the launch of the *Challenger* expedition in 1872, some naturalists complained that this unprecedented allocation of state funds would have been more usefully deployed to solve practical problems of the fisheries. For example, Frank Buckland (1826–1880), a naturalist and fisheries inspector, wrote, "The natural conditions of the bottom of this great North Sea is in a scientific sense less well known than the deserts of Sahara. Yet this 'Great Fish Farm' of Her Majesty's subjects is practically more important than the revelations made at vast expense to the country of the abyssal depths of far distant oceans."[136] Yet despite the lack of widely available scientific information concerning the oceans, it seemed clear to many others that improved marine science would help resolve practical problems of fisheries management. And indeed, when the London Fisheries Exposition opened in 1883, it provided a forum for scientific debate over the future of British fisheries.

Despite many warning signs, and ample fishermen's testimony, some scientists continued to claim that the oceans were not being exploited to their full potential.[137] Thomas Henry Huxley, in his address at the opening of the 1883 Fisheries Exposition in London, lamented that British fisheries had yet to undergo the industrial transformation that had already revolutionized commercial agriculture:

> Within the last quarter of a century, . . . agriculture has been completely revolutionized, partly by scientific investigations into the conditions under which domestic animals and cultivated plants thrive; and partly by the application of mechanical contrivances. . . . The same causes have produced such changes as have taken place in fishery, but progress has been much slower. . . . My astonishment was great when I discovered that the practical fishermen, as a rule, knew nothing whatever about fish, except the way to catch them. In answer to questions relating to the habits, the food, and the mode of propagation of fishes—points, be it observed, of fundamental importance in any attempt to regulate fishing rationally—I usually met with vague and often absurd guesses in the place of positive knowledge. . . . As to the application of machinery and of steam to fishery operations, it may be said that, in this country, a commencement has been made, but hardly more.[138]

In a later part of this speech Huxley then posed a question: "Are there any sea fisheries which are exhaustible, and, if so, are the circumstances of the case such that they can be efficiently protected?" He answered by stating his preference for deregulation: "A number of the most important sea fisheries, such as the cod fishery, the herring fishery, and the mackerel fishery, are inexhaustible."[139] Contrary to Huxley's claims, the plummeting of fish stocks was evident to many fishermen. The historian W. Jeffrey Bolster writes, "Fisheries science . . . barely existed. . . . Yet the 1850s and 1860s were noteworthy not only in New England and Atlantic Canada, but in Great Britain and Norway as well, for fishermen's insistence that governments do *something* to preserve the fish on which their livelihoods depended."[140] A scientist in attendance at meetings hosted by the Fisheries Exposition, Edwin Ray Lankester (1847–1929), countered Huxley's complacency and concurred on the need for authoritative scientific research. In a paper summarizing discussion at the meetings, he argued that it was "a mistake to suppose that the whole ocean is practically one vast storehouse":[141]

> If we are to have effective legislation at the present day in regard to our sea fisheries—we must, before proceeding any further, have *more knowledge*. Those . . . who earnestly desire additional restrictive Fishery Laws should do their utmost to enable zoologists to carry on researches which will provide that accurate knowledge of fishes and shell-fish, their food-reproduction and conditions of life—which must be obtained before legislation can reasonably be proposed.[142]

Lankester and his supporters recognized that commercial fisheries *were* in need of protective legislation. "More knowledge" was needed, but marine science was still in its infancy and lacked a strong institutional foundation. They advocated the construction of seaside laboratories for fisheries research. Citing established marine laboratories in France, Holland, Italy, and the United States, Lankester argued that "by offering free accommodation in such a laboratory to competent investigators you may obtain a large amount of valuable results at a minimum of expenditure. . . . Thus the working power and the general interest of the scientific world in these institutions and their work would be enormously increased."

Such sentiments eventually led to the creation of the Marine Biological Association and later to the construction of many British marine stations, including Plymouth, Lowestoft, and Millport.[143] In his own address at the Fisheries Exposition, Lankester offered his ideal layout for a marine station:

> Such a laboratory would stand near the shore, possess its own jetty and small harbour, with steam-launch for dredging and trawling, and other boats.... The basement... would consist of large well-paved rooms fitted with tanks, and an apparatus for the circulation of seawater. Here animals would be kept for observation, and the produce of a day's dredging or trawling would here be sifted and sorted. On the ground-floor and first-floor would be spacious rooms, with large windows... fitted with tables suited to the requirements of the microscopist. Small aquariums and pumping apparatus would also be provided.... A complete zoological and piscicultural library would be established.... The naturalists permanently and temporarily working here would in the course of a few years provide us with much-needed knowledge.[144]

Mounting fisheries concerns led to a reframing of marine biology as practical work in support of commerce and the state but also to efforts to define it as an exclusive professional domain in which amateurs had no place. In an article in *Nature* published in 1880 Lankester wrote, "That any enthusiastic young person who may unfold his umbrella on the seashore and contemplate under its shadow the starfish washed to his feet—should say that he has 'opened a zoological station' may be strictly true so far as the etymology of the words 'zoological' and 'station' respectively is concerned; but it is at the same time a misleading announcement, and likely to do more harm than good to the cause of zoological stations."[145] In Lankester's opinion, marine stations were crucial for the advancement of marine science; if they were to succeed, their purpose needed to be defined and standards of operation established. As we see in Chapter 2, the growth of marine stations in the latter half of the nineteenth century was instrumental in securing an institutional foundation for a science birthed in the field. Yet, dependent on government funding, these developments could not escape political embroilment exacerbated by increasing international competition over marine fisheries.

National competition over fisheries continued to foil international collaborative efforts in marine science well into the twentieth century. In 1899, a meeting was finally convened in Stockholm for organizing a collaborative survey and management plan for fish stocks in the North Sea. Eventually, these efforts led to the formation of the International Council for the Exploration of the Seas, but tensions remained. France refused to join until 1921. When the naturalist Edmond Perrier (1844–1921), president of the Muséum National d'Histoire Naturelle in Paris, addressed a gathering of the Congrès International d'Aquiculture et de Pêche, in June 1899, he concluded by noting that although the Stockholm meeting showed men of science eager to overcome "their nationalism for the greater good of humanity," such idealism was naïve. Notions of "perpetual peace and universal citizenship" were dangerous "utopias," he warned. "For war is the law of the world and competition the soul of social progress." Perrier called on fishermen to exercise their "vivifying action" on the navy and the country.[146] In his opinion—shared by other French naturalists smarting from their country's recent defeat in the Franco-Prussian War—the fate of France rested on her control of the marine environment; by supporting French fishermen, and the science of aquaculture, France might regain ascendency.

Conclusion

By the mid-1870s, new technologies allowed naturalists to explore previously inaccessible depths, a literate public in Europe and North America hungered for information about the sea, and a network of scientists increasingly identified themselves with a transnational scientific community working on the same problems. Time spent at sea, which Thomas Henry Huxley had once called "the worst & most unnatural . . . , fit for none but the unscrupulous," was now more likely to be cast as romantic, heroic, and even required for legitimate marine research.[147] The task of probing the aquatic environment was also becoming more standardized; the dredge, once merely a fishing tool, had been adapted as a scientific instrument.[148] With this framework in place, naturalists studying the ocean set out to expand the range and capability of their investigations.

In sum, the establishment of marine science cannot be appreciated without taking into consideration the wider context in which nineteenth-century society was turning its attention to the seas. Interest in both aquaria and fisheries expositions helped build public enthusiasm for ocean exploration. The industrialization of marine fisheries and fears for the long-term sustainability of fish stocks enmeshed commercial and military interests with those of scientists eager to carry out marine research. Government patronage, manifest in the supply of naval vessels for research, funding for the publication of scientific reports, and encouragement of international exchange for fisheries expositions was essential in establishing new foundations for cooperative expertise. Soon a new breed of vessels was created, designed and modified specifically to carry out deepwater oceanographic work.

Yet even as marine science was developing more standardized tools and procedures, the study of the oceans remained daunting and costly. The scale of observations required to map and understand problems of ocean circulation and of the global distribution of marine life demanded long-term observations and multiple expeditions. Oceanography was the "big science" of the nineteenth century, and it required the continued support of nation-states. As we see in Chapter 2, to secure long-term government support, the emergent science needed a firm institutional base. For a discipline born in the field, rather than in university or museum laboratories, establishing such a base was to prove challenging. Proponents of the new science of the sea soon learned to link appeals for funding to both nationalist and internationalist sentiment.

2

Marine Science for the Nation or for the World?

> I began the culture of oceanography, the new science which penetrates the secrets of the abyss. And this work has filled the best years of my life while absorbing the best of myself.... She has mitigated the sorrows which, little by little, take the place in the hearts of men first occupied by dreams of happiness.
>
> Albert I, prince of Monaco, *La carrière d'un navigateur*

> It may yet come to pass that when diplomacy fails—and it often comes perilously near failure—it will fall to the men of science and learning to preserve peace in the world.
>
> Arthur Schuster, "International Science"

The British *Challenger* expedition, the most oft-cited achievement of nineteenth-century marine science, is frequently honored as the key point of origin of modern oceanography. However, other developments during the late nineteenth century were of equal, maybe even greater, importance in establishing marine science. Taking these developments into account avoids the risk of seeing the institutionalization of oceanography as solely a British project or of overemphasizing the significance of particular expeditions by accepting at face value the cues left by organizers keen to memorialize their own achievements. Many more actors then emerge, and to find a useful framework for analysis we must decide which of them to examine in tandem. In this chapter I focus on developments in France, Italy, and Monaco. In these national contexts three figures loom large: the founder and promoter of early French marine stations, Henri Lacaze-Duthiers (1821–1901); Anton Dohrn (1840–1909), founder of the famous Naples Biological Laboratory; and the preeminent patron of turn-of-the-century oceanography, Prince Albert I of Monaco (1848–1922). All three sought to build lasting institutions, yet their motives were dissimilar. Their projects were shaped by very different perspectives on the future of international relations. Whereas Lacaze-Duthiers's and Dohrn's projects were tempered by nationalist loyalties, Albert's efforts to shape the direction of marine science were conceived in relation to his vision of building a foundation for world peace. Albert's project was motivated by his dual role as head of a small state and diplomatic intermediary between greater powers and his belief that the marine sciences were uniquely suited to advancing humanitarian goals.

A Context for Motivation: The Marine Station Movement

The success of the *Challenger* expedition set a new standard for oceanographic work that scientists in other countries were keen to emulate.[1] Persistent lobbying by a retired naval officer and amateur conchologist, the marquis Léopold de Folin, eventually resulted in four state-sponsored expeditions by France.[2] The voyages of the *Talisman* and *Travailleur* (1880–1883) were the government's response to the *Challenger* expedition (see Figure 2.1).[3] Although successful in gathering a few new benthic species, these expeditions fell far short of *Challenger*'s achievement.[4]

Ultimately, efforts to establish a sustained state-sponsored program for deep-sea exploration failed to take hold in France.[5] The absence of a deep-water program—particularly when compared to the Coast and Geodetic Survey and Fish Commission programs in the United States—revealed a notable lacuna in France's scientific agenda. On the other hand, France excelled at developing a coastal research program.

Between 1840 and 1900 marine stations were built throughout Europe and North America and as far afield as Japan.[6] In 1923, William Herdman, one of the first historians of oceanography, wrote of marine biological stations, "Fifty years ago the biological station was almost unknown; now there are, I suppose, about fifty or possibly more, large and small, scattered along the shores of the civilized world from the Arctic Circle to the tropics and Australia, from western California to far Japan in the East."[7] In his view, marine stations returned naturalists, too long immured in

Figure 2.1 Imaginative illustration of the discoveries of the *Talisman* and *Travailleur*. Jules Gros, "Explorations sous-marines du 'Travailleur' et du 'Talisman,'" *Journal des Voyages*, no. 409 (May 10, 1885): 296–297.

university laboratories, to contact with the natural world. But in reality the birth of marine stations was closely tied to the rise of the university laboratory and the formalization of university-level science instruction. Naturalists were not so much escaping academic laboratories, as laboratories were escaping the confines of university campuses. As part of a growing movement in the biological sciences focused on place-based research and the study of living organisms, and away from museum collection study, marine stations provided a hybrid middle ground between laboratory and field. They offered a space where living creatures could be collected and examined under controlled conditions, thus maintaining the authority of the laboratory at a distance from the research centers of the metropole.[8]

The marine station movement can be understood only within a context of international rivalries, the widespread exploitation of marine resources, and a changing transnational culture of science. These were sites of projects to popularize science, enhance national prestige, and democratize access to education.[9] Despite being situated within national boundaries, nineteenth-century marine stations permitted unprecedented international cooperation and presaged the internationalist movements that shaped marine science in the twentieth century.

By some accounts modern marine zoology began on the coast of Normandy, where Georges Cuvier (1769–1832), the father of paleontology, spent the years of the French Revolution's Reign of Terror studying mollusks, but France is often relegated to passing comment in English-language scholarship on the history of marine science (focusing exclusively on the French *Talisman* and *Travailleur* expeditions in the 1880s or, jumping ahead, on Jacques Cousteau's technological innovations in the twentieth century).[10] There are multiple possible explanations for this omission, some of which I have addressed in greater detail elsewhere.[11] It is clear, however, that in terms of the sheer number of marine stations established along its coasts France was unmatched. Twelve stations were built between 1840 and 1900.[12]

The first French marine station was established in 1859 by the embryologist Victor Coste (1807–1873) in the town of Concarneau on the south coast of Brittany. Coste described his station as an "observatory of the ocean" where all arrangements had been made according to the requirements of "advanced studies."[13] However, critics derided the station as little more than a public aquarium.[14] Following the station at Concarneau, two more privately funded and operated stations were founded in France, one at Arcachon in 1867 and another at Sébastopol in 1871. The best known, the Roscoff marine station, was established in 1872. Anton Dohrn's celebrated Naples, Italy, station was opened with Prussian backing in the same year. A Belgian report of 1897, while acknowledging it as a "palace installed in the largest and most luxurious fashion," was quick to point out that the Roscoff station was the "first true university station, that is to say dedicated to the practical research of students, and not solely to the studies of professors."[15]

Linking the Metropole to the Coast

The Roscoff station was the project of Henri Lacaze-Duthiers, the single most influential figure in the construction of marine stations in France. Born to an aristocratic family in 1821, the young Baron Lacaze-Duthiers studied medicine but soon discovered a passion for marine biology. By 1854 he was a professor of zoology at the University of Lille, where he received government funding for a voyage to the Mediterranean to study corals along the coast of Algeria.[16] The resulting *Monographie du Corail*, acclaimed both in France and abroad, secured his reputation as a specialist in coastal marine biology.[17] He capitalized on this initial success by relocating to Paris in 1864 to work at the École Normale Supérieure, subsequently becoming professor of zoology at the Museum of Natural History at the Jardin des Plantes in 1865, professor at the Sorbonne in 1868, and member of the Academy of Sciences in 1871. The accomplishment for which he is best remembered is his establishment of stations at Roscoff in 1872 and at Banyuls-sur-Mer in 1881.[18]

Initially, Lacaze-Duthiers's vision of the future of French marine biology extended much further than the tidal flats of Roscoff. His intention was to begin "a survey of the zoological riches of the coasts of France, and to direct youths, and zealous adepts in zoology" in research similar to that long pursued by the English.[19] So envisaged, coastal marine biology was to serve the national interest, revealing the biological "riches" of France. By midcentury, British naturalists possessed unrivaled technological expertise in collecting marine specimens at depth (paralleling British expertise in submarine telegraphy) and were also developing their own coastal stations.[20] Lacaze-Duthiers, aware of the coastal surveys already completed by the British, believed that marine science had fallen behind in France. Roscoff would be a starting point, serving as a baseline for the comparison of biological diversity elsewhere along the French coast.[21]

France's Ministry of Public Instruction provided a financial subsidy of 3000 francs per year for the maintenance of the Roscoff station. Yet funding often fell short, and Lacaze-Duthiers was forced to cover expenses of the station himself.[22] As the historian Harry Paul has noted, "Fortunately, Henri Lacaze-Duthiers, an old bachelor content with the sexual activity of marine life, could afford her."[23] The Roscoff station was gradually improved with purchases and donations of scientific equipment, including a boat, the *Dentalium,* a gift of the French Association for the Advancement of Science.

One of the earliest private benefactors was the Belgian-born algologist Jean Chalon. In 1898 he published a short review of his work at Roscoff, highlighting the many advantages of the region for marine botanical work in the bulletin of the Belgian Royal Botanical Society. He remarked that the library contained a beautiful herbarium of the plants and algae of Brittany. Like other visitors, he also noted the uniqueness of this provincial part of France whose inhabitants spoke Breton and often passable English and where the French language had only recently come to be taught in elementary school. One still encountered elderly men, and certainly elderly women, "who [knew] not a single word of French."[24]

Parisians traveling to Roscoff in the late nineteenth century might well have felt themselves in a foreign country. Linguistic differences catalyzed distrust and hostility. Unlike the rest of France, which was primarily agricultural, Brittany's economy depended on a few commercial oceanic fish species, notably sardines. The historian Eugen Weber has argued that, around 1870, what unity France had was "less cultural than administrative." Only at the very end of the nineteenth century, through the shared experience provided by "roads, railroads, schools, markets, military service, and the circulation of money, goods, and printed matter," did French culture begin to become truly national.[25]

French-speaking travelers sometimes compared the Bretons to "savages," seeing the improvements to be brought to the region as moral as well as technological progress.[26] As the rector of the Academy of Rennes advised in a report on the Breton departments in 1880, "Frenchify Brittany as promptly as possible . . . ; integrate western Brittany with the rest of France."[27] Marine stations dispersed along the coast and administratively connected to the metropole served as nodes linking such peripheral areas to urban-based power centers.

Reinvigorating the State through Science

A record of the early years of the Roscoff station is provided by Henri Lacaze-Duthiers in the *Archives de zoologie expérimentale et générale*, a journal he founded and whose first issue, slated to appear in July 1870, was delayed for two years because of the Franco-Prussian War. Lacaze-Duthiers, like many other French scientists demoralized by the war, believed that the reinvigoration of French science was necessary for rebuilding the power and prestige of the nation. Listing his many Prussian collaborators from the years leading up to the war, "those who science had once made friends," he grumbled that he had not received a single word of condolence when the ancient Jardin des Plantes (the scientific heart of France) was destroyed by Prussian bombs.[28] In his introduction to his new journal he observed, "The awakening of the intellectual movement in France is to our eyes inevitable. Its rationale is in our defeat. It must be without limit like our disasters and our misfortune."[29] Lacaze-Duthiers insisted that he never allowed politics to interfere with his scientific work. Nevertheless, in his view, rejuvenation of French science must not rely on those who had revealed themselves as enemies of scientific internationalism. Lacaze-Duthiers approached the establishment of marine stations as part of a project of national restoration and retained a grudge against Prussian scientific colleagues, but he still viewed science as a universal endeavor that elevated the human spirit.[30]

The *Times* of London published a short article summarizing work at the Roscoff marine station on March 3, 1881.[31] The author reported "remarkable progress," stating that what had at first been "merely a primitive aquarium under a shed at the side of a garden" had become a "well-equipped station." By 1883 an overnight train connected Roscoff with Paris. Students and professors traveled seasonally between Paris and the Roscoff and Banyuls stations, spending part of the summer at Roscoff and visiting Banyuls, on the Mediterranean, in winter.[32] Lacaze-Duthiers, convinced that all naturalists needed direct, hands-on experience with their subject of study, declared that the purpose of both stations was "to make naturalists."[33]

As far as he was concerned, the Prussian-backed Naples station would never accomplish this aim because it removed students from the collection process, and unlike the French stations, it failed to serve educational purposes. He recalled how a foreign naturalist, having previously worked at Naples, asked him for permission to finish some embryological research at Banyuls. "You will find these animals when you look for them down below the laboratory . . . in the same conditions in which you collected them at Naples," Lacaze-Duthiers informed him. But the young naturalist replied, "I have never seen them in situ; they were brought to me, and I do not know their habitat."[34] As Lacaze-Duthiers explained in one of his reports, "The young naturalist going to Roscoff will find all the means necessary for his work, but he will be forced to go out on the tidal flats, and this is above all what I desire, to look himself for animals and thus form his education."[35] Exemplary of the sort of experience Lacaze-Duthiers desired for researchers was that of twenty-seven-year-old Léon Frédéricq, who worked at the Roscoff station for six weeks during the summer of 1878. By telegraph he wrote to his parents, "Very little time, am dissecting octopuses, tormenting octopuses, feeding octopuses, caressing octopuses, watching over octopuses, educating octopuses, dreaming octopuses, conversation only octopuses."[36]

The use of marine stations for research and instruction emulated a long-standing French scientific institutional model. The preeminent scientific research institution in France, the Museum of Natural History in the Jardin des Plantes in Paris, had offered training courses in zoological fieldwork as early as 1815.[37] In this tradition the marine stations were field schools where visiting students could learn, free of charge, the practical skills required to pass their university exams and be offered space and resources to work on thesis projects. The Roscoff and Banyuls model, combining research with instruction, set a precedent for other French stations and was emulated by Lacaze-Duthiers's students: Alfred Giard at Wimereux in 1874 and Edmond Perrier at Tatihou in 1887.[38]

Little in the organization of the station forced any particular specialty on visiting students. This freedom was appreciated, as was the social world of the station. Entries in the visitor log praise the sense of community it fostered. Describing recent expansions, one visiting naturalist enthused, "It is a scientific city that has developed on our coast of Brittany, just as in those large American cities which in but a few years replace a simple hamlet."[39] Another visitor described "the most agreeable camaraderie." At Roscoff he had found himself "in a real scientific circle, a little world full of work and enthusiasm."[40]

Lacaze-Duthiers expressed pride that his stations permitted students and naturalists with limited means access to specimens and tools that would otherwise have been prohibitively expensive, and he criticized Dohrn's Naples station as accessible only to those who could pay (but he never questioned the quality of the scientific output). Also offensive to Lacaze-Duthiers's scientific liberalism was Dohrn's insistence that the Naples station specimen preservation methods be guarded as trade secrets.[41] Although the Naples station was fundamentally different in its organization, Lacaze-Duthiers's criticism had more to do with politics than with practice. The Naples station received significant financial backing from Prussia, and Anton Dohrn proudly identified as Prussian.

Unfortunately, Lacaze-Duthiers's hostility contributed to the increasing insularity of French marine scientists, which had long-lasting effects on the development of oceanography in France. Physical oceanography became the dominant branch of marine science in the rest of Europe, but marine zoology continued to hold center stage in France.[42] There were also financial reasons why French scientists did not visit the Naples station. Reserved visiting research positions ("tables") were rented by governments or institutions, and the French government was unwilling to pay the requisite fee—justifying this decision by pointing to the abundance of alternative coastal stations in France.[43]

Anton Dohrn and the Naples Zoological Station

The Naples station was founded, like that at Roscoff, in 1872. Dohrn was tireless in his efforts to promote this station, which he hoped would become part of a network of stations under his direction.[44] Born to a wealthy family in the East Prussian city of Stettin (now Szczecin) in 1840, Dohrn at first embarked on a military career.[45] Like his father, an amateur naturalist, Dohrn had other passions and soon abandoned the military. He entered the University of Jena in 1862 where, studying with Ernst Haeckel, he was introduced to Darwinism. Under Haeckel's tutelage, Dohrn became increasingly interested in the study of marine organisms, a project that he took up with characteristic energy.[46]

Naples was not Dohrn's first choice of location. Initially, he had hoped to establish a station on the Strait of Messina, between Sicily and the Italian peninsula, favoring this region because of local oceanographic conditions: upwelling facilitated the collection of marine organisms.[47] In a letter to Darwin, Dohrn laid out his plans:

> Having stayed now several times on the seashore for zoological studies, I have found how difficult it is to study Embryology without an Aquarium. This want has suggested to me the idea of founding not only Aquariums, but Zoological Stations or Laboratories on different points of our European coast. Such a Station should consist of a little house of perhaps four rooms, an Aquarium connected with the sea and the house,—the Aquarium of perhaps 60 feet in Cubus, where one might have streaming water,—a boat for dredging work, dredges, nets, ropes,—in short, all that is necessary for a marine zoologist. . . . All this might be had at a not too high price at Messina, where I thought of founding the first Station.[48]

Ultimately, Naples proved more affordable, although Dohrn continued to complain that the rain, heat, and mosquitoes (and the consequent risk of malaria) dissuaded would-be visitors during the summer.[49] Nevertheless, a booming—and heavily Prussian—tourism industry guaranteed a source of revenue for the station through the sale of tickets for an attached public aquarium.[50] Naples had long been a favored destination for German scholars and scientists following in the path of Goethe.[51] But working in Naples instead of Messina also meant working in a less ideal environment. The Bay of Naples was polluted, and water pumped into the aquarium first had to be filtered.[52] Also, Dohrn, like Lacaze-Duthiers, had conflicts with the local population.[53] As he wrote in a letter to Darwin, "The difficulties in this country are something quite unheard of for all of us northern people. The indolence, dishonesty, hatred even against a good and disinterested enterprise, are quite regular qualities with this people, and it wants one's last resources of nervous energy to overcome the physical hindrance and the moral disgust, it fills one with."[54] Dohrn (like many other Continental naturalists) turned to the British for custom-order material and technological expertise. All the equipment for the Naples laboratory had to be imported, including pumps, pipes, and sheet glass—much of it from England.[55]

Yet despite initial challenges, the Naples station thrived, attracting scientists from many countries and fostering international scientific alliances. Dohrn, a master promoter with boundless energy, secured support from influential naturalists throughout Europe. To work at Naples became a rite of passage. Soon it was the norm to refer to the station as a "Mecca."[56] In his report on the biological stations of Europe, the American naturalist Charles Kofoid lauded the Naples Station as preeminent, "inspiring in its history and unparalleled in its growth, unsurpassed in its contributions to biological science, profound in its influence upon the course of development of modern biology, and powerful in its stimulus to the establishment of biological stations elsewhere."[57] He even attributed the scarcity of marine stations in Italy compared to the many in France to the Naples station's overwhelming success.[58] Certainly naturalists who came to work at Naples were not disappointed by what they found. For instance, the Norwegian explorer Fridtjof Nansen, who visited in 1886, described the station as a "central organ for zoology ... where research is carried on with assiduity, and where the burning scientific questions of the day are sifted and discussed in a fashion which helps in no small degree to render a stay ... inspiring and profitable."[59]

Unlike Roscoff and Banyuls, where resources were primarily directed toward instruction, Naples was promoted first and foremost as a private research institution. Station regulations stipulated that visiting scientists were not allowed to preserve their own specimens; this would be done for them "upon request" and "for a reasonable price."⁶⁰ Much of the science conducted was based on the experimental work of the German embryologist Wilhelm Roux (1850–1925). Sea urchin embryos were subjected to centrifugation, exposure to chemicals, and electric and magnetic fields.⁶¹ The purpose of these experiments was to determine whether development was driven by external stimuli or internal mechanisms.⁶²

Dohrn and Lacaze-Duthiers, whose scientific projects often ran in parallel and whose geographic terrains did not overlap, nevertheless cast one another as competitors. Lacaze-Duthiers referred to Dohrn as "that Prussian scientist in Naples," and Dohrn referred to Lacaze-Duthiers as his "particular rival."⁶³ In a letter to his sister, Dohrn mocked Lacaze-Duthiers, claiming that he had threatened to destroy his station at Roscoff and resign his position from the Sorbonne should the French government decide to fund a table at the Naples laboratory.⁶⁴ The first French naturalist to work for an extended period at the Naples station was the biologist Maurice Caullery in 1906—five years after Lacaze-Duthiers's death. He did so with the encouragement of Alfred Giard, the founder of the Wimereux marine station and a former student of Lacaze-Duthiers, while acknowledging that the long French abstention from Naples, and isolation from the international science it hosted, was the result of the station's association with Germany.⁶⁵

Lacaze-Duthiers and Dohrn may have been fiercely competitive, yet both men favored opportunities for engagement with marine work abroad. For example, the Egyptian oceanographer and historian Selim A. Morcos has shown that the Alexandrian naturalist Osman Ghaleb, who studied with Lacaze-Duthiers at Roscoff, maintained a lengthy correspondence with him after his return to Egypt. Lacaze-Duthiers wrote approvingly of his pupil, "[Ghaleb] spent one month at Roscoff to familiarize himself with the methods of fishing and the research applied in the station, with a view of the studies he is planning to carry out in the Red Sea."⁶⁶

The Paradox of French Marine Science

In France, institutional linkages existed between the navy and the fishery, between the fishery and coastal marine stations, and between marine stations and universities. These institutional networks were strengthened by a culture that valued the democratization of science education: a tradition stemming from the French revolutionary ideals of *égalité* and *fraternité*. All these factors bolstered public support in favor of marine biology, which came to take precedence over more quantitatively based physical oceanography.

For all European countries, fisheries research required international political agreements because national fisheries extended into the extraterritorial high seas. Political friction in the North Sea caused by fisheries competition led, in 1882, to the signing of the North Sea Convention by Britain, Belgium, Denmark, France, Germany, and the Netherlands. The signatories relinquished territorial jurisdiction over shared fishing grounds, setting in place the groundwork for a program of international fisheries management.[67]

The International Council for the Exploration of the Seas (ICES) held its inaugural meeting on July 22, 1902, in Copenhagen. Denmark, Finland, Germany, the Netherlands, Norway, Sweden, Russia, and the United Kingdom participated, but France was notably absent. The Scandinavian countries, especially Norway and Sweden, reliant on the North Sea fishery, were the strongest supporters of the new program. In the decades leading up to the agreement, Scandinavian fisheries fluctuated wildly, resulting in famines, and the threat of stock collapse did much to encourage government support for fisheries research.[68] The ICES scientific program—shaped by concern for the fisheries—focused primarily on discerning the physical properties of the oceans (salinity, chemistry, currents, and meteorology) rather than on the more traditional concerns of classification, physiology, and morphology of organisms.[69] As a result, physical oceanography began to emerge as a distinct branch of marine science, more reliant on quantitative modeling than was marine biology.

Many French naturalists campaigned for the inclusion of France in the newly formed international program.[70] However, the decision to keep France out of ICES seems to have been determined not by the scientific community but by the French government. Théophile Delcassé, director of the French merchant marine, submitted a report to parliament warning that French participation in ICES would impose a considerable burden on the state. The consensus among parliamentary officials was that, in view of the funding already allocated for fisheries research, further spending would be redundant.[71] French naturalists in favor of ICES saw their country's reluctance to join as an affront both to the international scientific community and to the young discipline of oceanography.[72] Thus, in a 1903 letter to the Swedish oceanographer Otto Pettersson (1848–1941), the French oceanographer Julien Thoulet wrote,

> I do not have a very favorable response to give you regarding your request for the collaboration of vessels of the French state. Despite incessant efforts, I have not been able, for the past twenty years that I've been working, to shake the French government out of its indifference or, better said, out of its hostility to oceanography.[73]

French scientists interested in physical oceanography, foremost among them Julien Thoulet, found themselves out of step with the national program. As a result, Thoulet had to seek support for his work with deferral to biology, arguing that physical oceanography (which he called simply oceanography) and marine biology (which he called zoology) were complementary disciplines because both were of practical importance to the fishery.[74] His views on the foundational importance of physical oceanography for marine biology are apparent in his book *The Ocean: Its Laws and Its Problems*:

> We will not understand [the distribution of marine life] until naturalists who study marine animals ... have adequately enlightened themselves through oceanography. The point of this is not to establish a preeminence between the two sciences: oceanography and zoology, but there is for one, oceanography, anteriority, because of its application to the other, zoology.[75]

In the same text, Thoulet extends this order of application to the fishery, suggesting that the thermometer can become a veritable fishing implement.[76] Despite his best efforts, Thoulet's appeals fell on deaf ears. But looking beyond France for support, he found a powerful ally, one whose own ideas about the future of oceanography were more accommodating to open-ocean scientific investigations. Thus, Thoulet came to place all his hopes for the future of French oceanography in Prince Albert I of Monaco, writing to Pettersson, "You have been in a position to appreciate how great and sincere is his devotion to the science of oceanography.... I must admit to you, between us, I have much more hope in [him] than in the real or effective cooperation of the French government."[77]

The Prince of Ocean Science

Prince Albert completed twenty-eight oceanographic cruises aboard his yachts, modified for scientific work, over the course of his lifetime and often used oceanographic instruments of his own design. He corresponded with leading marine scientists in Europe and the United States, and he founded a research institute in Paris and an oceanographic museum in Monaco. He was a major patron of marine science—supporting marine stations in France, funding the development of new instruments, and undertaking initiatives to facilitate international scientific cooperation and technological exchange (see Figure 2.2).

Albert's political and dynastic status as reigning sovereign of the tiny principality of Monaco set him apart from other naturalists engaged in marine work. His financial resources and public stature facilitated his

Figure 2.2 Prince Albert I at work in his cabin aboard one of his yachts. Albert, prince of Monaco, *La carrière d'un navigateur* (Paris: Librairie Hachette et Cie, 1913).

success, and his celebrity as both naturalist and patron of science dignified his international standing as head of state. Albert sought to embody an ideal of the enlightened monarch, as a patron of higher learning and champion of international peace and social justice. He insisted that the promotion of science illuminated a path to good governance, claiming, "I cultivated science because she spreads light and light engenders justice, the guide without which a people marches toward anarchy and decadence."[78]

Whereas other important figures in the early history of oceanography—naturalists in France, England, and the United States—struggled to secure state support, Prince Albert *was* the state. As the ruler of a small principality surrounded by French territory, he had to operate within an overarching French political sphere of influence, yet he did so with some autonomy. For instance, he voiced his support for the beleaguered French officer Alfred Dreyfus, earning the rancor of the French elite.[79] As a result of Monaco's geographic enclosure within French territory, the history of Albert's oceanographic work is inseparable from the history of the development of marine science in France.

Born in Paris in 1848, Albert spent his youth outside Monaco, attending school in Paris and Orléans before beginning two years of naval service at the age of seventeen, first in the Imperial French Navy and then in the Royal Spanish Navy. While new discoveries were revolutionizing the marine sciences, political and military events were dramatically altering Prince Albert's future as a head of state. The Grimaldi family has ruled Monaco since 1297; their centuries-long survival required astute political negotiation with powerful neighboring nations. The townships of Menton and Roquebrune, previously ruled by the princes of Monaco, were ceded to the kingdom of Sardinia in 1848, the year of Albert's birth, reducing the principality to a single commune. Sardinia was annexed by France during the Italian Wars of Independence, and after 1860 France controlled all the territory surrounding the principality. Albert's father, Charles III, secured the financial stability of the tiny state by developing Monte Carlo as a gambling haven.[80]

Although he was a prince, Albert's political power on the international stage was insignificant. But his reign was domestically transformative; when he inherited the throne in 1889, he set about modernizing the principality.[81] He reformed the education system, established new schools, created professional institutes, and built a public library. In 1910 he declared the election of a parliament through universal suffrage, effectively ending the absolute rule of the Grimaldi family.[82] Although these reforms may have been instituted in part as a response to public pressure, Albert aspired to embody the ideal of an enlightened monarch, a strong counter to the dark shadow the Monte Carlo casino threatened to throw on the moral prestige of the principality and its prince. It was above all through marine science that Albert aspired to exercise influence beyond the confines of his small kingdom.

Albert spent much of his time away from Monaco, often traveling to Paris, where he led a princely social life and attended scientific lectures. It was also in Paris that Albert discovered oceanography. By his own account, the 1884 exhibit at the Museum of Natural History of the collections brought back by the *Talisman* and *Travailleur* first sparked his interest in oceans (see Figure 2.3). He recalled, "I became convinced that I could do better in this field of study than all my predecessors, and I put myself to work despite a complete lack of encouragement from my immediate entourage."[83] Albert nevertheless did find encouragement in Parisian scientific circles. He soon forged a close relationship with the professors of the Museum of Natural History and befriended Alphonse Milne-Edwards.[84] The scientific world opened avenues to international honor and recognition, from which he was otherwise excluded.

The majority of Albert's scientific projects were devoted to oceanography, but he was also interested in paleontology. In 1902 he built a museum of prehistoric anthropology in Monaco, followed in 1910 by an Institute of Paleontology in Paris. Despite oceanography and paleontology being very different fields of study, Albert understood both as working toward a shared goal: human progress. The study of oceanography revealed the origins of biological life, whereas paleontology revealed the progressive path of human evolution. In his inauguration speech at the opening of the Institute of Paleontology—which he described as a "temple"—Albert explained, "In bringing together the history of the ocean and [the history] of Life, I am but respecting a law of modern science that prepares in the progressive fusion of its elements a magnificent domain for coordination of human work in our enlarged

Figure 2.3 Exhibit of the collections of the *Travailleur* and *Talisman* at the Paris Museum of Natural History. "Guide a l'exposition sous-marine du Travailleur et du Talisman" (Paris: G. Masson, 1884), 8–9.

brain."[85] The theme of science as an outcome to evolution and a guiding light to a better future recurs in many of his speeches.

Prince Albert set off on his first lengthy oceanographic cruise in the summer of 1885 aboard his yacht *Hirondelle*.[86] This would be the first of four consecutive summers of oceanographic work and research in the North Atlantic. The *Hirondelle*, a wooden schooner built in 1862 in a shipyard in England, lacked onboard laboratories and refrigerated storage but was fitted out with a deep-sea trawl similar to that used aboard the US survey vessel *Blake*.[87] With it, Albert sounded and dredged in the Gulf of Gascony and on the Grand Banks. He also conducted drift experiments to study the Gulf Stream, launching 1700 floats (224 of which were recovered) over a period of three years.[88] As Albert later wrote in his autobiography, "I was a sailor, and my passion for science permitted me to recognize what useful work needed to be done; ... my plans were dependent only on me, and I could modify them in course of the voyage following changes of circumstance. In a word, I centralized in my head ideas and will."[89]

By 1889, Albert had amassed enough material from cruises in the Mediterranean and Atlantic to put his collections on display at the Universal Exposition in Paris. These oceanographic displays took up half of Monaco's allotted exhibit space, and the pavilion became a popular attraction.[90] In 1891 he acquired a new custom-built yacht, christened the *Princesse Alice* after his second wife.[91] The new vessel, a three-masted schooner, was state of the art, complete with an electric dynamo, a powerful engine, refrigerated compartments, three onboard laboratories outfitted with distilled and saltwater circulation systems, and gimbaled tables.[92]

In 1891, in a speech to the Royal Society of Edinburgh titled "A New Ship for the Study of the Sea," Albert diplomatically flattered his audience, praising the British *Challenger* expedition, but also criticized the national expeditionary model of research as outdated and inefficient. Nations should not carry out expeditions, he argued, because nations were fundamentally militaristic and would always favor military over scientific expenditures.[93] By contrast, he claimed, his own expeditions were not the project of a state, but of one man. Marine science was attractive to Albert because the geographic scope of such research, on the high seas beyond national jurisdiction, necessitated international collaboration and encouraged political and military accord. Albert's internationalist ideals were reflected in the makeup of the scientific crew he invited to join him.[94] One of the most influential of his scientific collaborators was the British naturalist John Young Buchanan (1844–1925), chemist of the *Challenger* expedition. Buchanan took part in Prince Albert's summer expeditions of 1892, 1894, and 1898 aboard the *Princesse Alice*.[95]

Prince Albert's third vessel, *Princesse Alice II*, was launched in 1897.[96] And with this new ship he ventured as far afield as Spitsbergen (in the Svalbard archipelago) in the summers of 1898 and 1899.[97] Albert described his expeditions as beneficial for humanity and as examples of the benefits of international collaboration. The members of the 1898 expedition to Svalbard included representatives from France, Germany, Italy, the United Kingdom, and Norway. Having sought to modernize his principality in his role as monarch, Albert took it on himself to direct, improve, and promote oceanographic science not only beyond its borders but beyond the social world of scientific specialists. He saw the international popularization of science as key to the success of his project. As Buchanan wrote to him, after a notice of the prince's scientific work had appeared in a London newspaper,

> It was exactly what was wanted for the information of the British public, of the serious work which is being done at Monaco and having appeared in the *Times* it is the duty of every British subject to know it, or, at any rate, to pretend to know it. A hundred communications to scientific periodicals would not have the same effect, and the genuine sporting spirit which pervaded, naturally appealed to the many who care little for science.[98]

The ambitious research program Albert envisaged required a dedicated staff, and Albert assembled an ever-growing number of specialists possessing particular complementary forms of expertise in marine work. By 1887 he had engaged the French naturalist Jules Richard (1863–1945) in his service. Richard served as secretary of scientific work, curator of scientific collections, and head of the laboratory aboard the prince's yacht.[99]

New Institutions for Marine Research

Conceived by Albert as a "temple dedicated to science and art," the Musée Océanographique clings to a cliff overlooking the Mediterranean. At its inauguration ceremony on March 29, 1910, Prince Albert declared, "Here, ... out of Monaco's earth has sprung a proud and inviolable temple, dedicated to the new divinity that reigns over the best minds. I have lent the forces of my brain, my conscience, and my sovereignty to the extension of scientific truth, the sole terrain on which the elements of a stable civilization can grow."[100]

One of the principal functions of the museum was to collect existing oceanographic instrumentation and facilitate the creation of new designs.[101] The museum also coordinated international research programs. The architecture of the new building awed visitors and, as an emblem of the principality, rivaled Monte Carlo. Its ornate facade monumentalized pioneering oceanographic voyages.[102] The museum was also a station for local marine studies, and a twenty-ton steam vessel, the

Eider, was assigned for this task.[103] Today, the main entrance is flanked by two allegorical sculptures: *Truth,* unveiling the forces of the world to science, and *Progress,* coming to the aid of humanity. Mosaic floors depict the *Princesse Alice* and sea creatures; light fixtures are shaped to resemble medusae and radiolarians, and sculptures and stained glass with marine themes adorn a monumental central staircase. The museum is divided into three great exhibition halls: for physical oceanography, applied oceanography, and zoological oceanography—understood as the three branches of marine science (see Figure 2.4).[104]

Describing the opening ceremonies, the British journal *Nature* reported, "The inauguration was an arresting function, which could not fail to impress the most regardless pleasure-seeker . . . with the thought that science, even in, perhaps, its least known department, was a thing of high importance." The author regretted the absence of any British representatives, "a cause of humiliation" and a "slight upon a noble enterprise."[105] The British absence was noteworthy in light of tense diplomatic relations between attending parties, France and Germany, who both, unlike Britain, relied on Prince Albert as a diplomatic intermediary.

Figure 2.4 The hall of applied oceanography of the Musée Océanographique. From a contemporary postcard; no date is given.

Albert's role as a political negotiator was necessary for the principality's long-term survival. As he explained in a public address in 1909, "Only one foreign policy is possible for our country: it is even the policy necessary to defend our vital interests in the midst of European instability. And it implies a role your prince must fill by acquiring, through moral force of character, the influence others obtain through force of arms."[106] With Albert's sponsorship, Monaco hosted the eleventh Universal Peace Congress in 1902 (the vast majority of attendees were, however, French). The following year he established an International Peace Institute with the declared aim of producing scholarship on international law, statistics of war, and the development of international institutions, all with the larger aim of promoting pacifism.[107]

Albert's internationalist efforts extended to all aspects of his oceanographic work. He planned an international oceanographic congress to convene in Monaco in 1906. Although the meeting never went ahead, correspondence related to its planning indicates the extent of international support for Albert's project.[108] Particularly revealing is a letter from Otto Pettersson:

> Monaco fulfills most conditions for becoming the center of an international investigation of the sea, being independent of political changes and opinions of legislative bodies as well as the scientific rivalry of the universities and academies of the greater nations. The oceanographic museum of Monaco, if combined with an international investigation of the Atlantic and other oceans, might become the seat of this organization and the literary center for publishing the general results.[109]

Completion of the museum corresponded with a period of growing civil unrest in Monaco, culminating in the Monegasque Revolution of 1910.[110] But even so, the period between 1905 and 1910 saw the height of Albert's scientific activity. Albert undertook regular summer research cruises, traversing the North Atlantic from the Azores to Spitzbergen. Perhaps further irritating his subjects, Albert's ambitions as a scientific patron had by now led him to invest resources and direct his energies well beyond the confines of Monaco.

In 1911, Albert inaugurated his Institut Océanographique—a permanent institution for oceanographic instruction in the heart of Paris. Sometimes described as an oceanographic school, the institute, which shared an administrative council with the museum, was designed to promote oceanography through university courses offered in conjunction with the Sorbonne, to publish scientific papers, and to host public lectures.[111] A report in the journal *Science* described the new institution as "at once French and international in character."[112] An administrative committee of scientists included John Murray (British), John Young Buchanan (British), Erich Dagobert von Drygalski (Prussian), and Fridtjof Nansen (Norwegian).[113] A secondary administrative council, entirely French, included Yves Delage (director of the Roscoff marine station), Paul Fabre-Domergue (former assistant director at the marine station of Concarneau and general inspector of fisheries), Louis Joubin (a veteran of the *Travailleur* and *Talisman* expeditions and former director of both the Banyuls and Roscoff stations), Edmond Perrier (director of the Paris Museum of Natural History, veteran of the *Talisman* and *Travailleur* expeditions, and founder of the Tatihou station), Paul Portier (who had accompanied the prince on a voyage to Spitsbergen), Jules Richard, and finally, Julien Thoulet, who years earlier had placed his hopes for the future of French oceanography in Prince Albert.

The institute provided a venue for free public lectures and slide shows on oceanography, and word of their popularity was reported as far away as San Francisco. As the author of a 1913 article in the *San Francisco Chronicle* wrote,

> In a most magnificent lectures hall, addresses on the deep sea and its inhabitants are now riveting the attention of studious and fascinated audiences.... The magic lantern and the cinematograph are pressed into the lecturer's service.... The lecturers are mostly from the Sorbonne, but the most popular lecturer of them all is Prince Albert of Monaco, who just now is one of the idols of Paris. Whenever this princely oceanographist is announced for a lecture, the poorer holders of tickets sell their right of admission for a handsome price. And the demand is enormous.[114]

Prince Albert succeeded in capturing the public imagination, but did he succeed in redirecting French government support toward marine research? An important source documenting the origin of the Oceanographic Institute survives in the form of a letter from Albert to the French minister of public instruction dated April 25, 1906:

> Having devoted my life to the study of Oceanographic Sciences I have been struck by the importance of their action on several branches of human activity, and I have striven to obtain for them that place which they should occupy in the solicitude of governments.... Several States have already organized scientific cruises ... and have established a solid basis for the development of Oceanography; but France, notwithstanding that the science of the sea presents for her special interest, has not treated it with the same liberality as she has treated other branches of Science. Nevertheless, for some years past I have caused to be given in Paris a series of lectures which have been followed by audiences each time more numerous and more attentive.... I then wished to fill a gap by myself creating and establishing in Paris a center of Oceanographic Study closely connected with the laboratories and collections of the Oceanographic Museum at Monaco.... Desirous that this institution shall survive me..., I beg the French Government to recognize it as of public utility and to approve its Statutes.[115]

From a report sent to Albert in September 1910 we learn that he had begun to exert his influence on the marine station at Roscoff. There is no evidence that the prince ever conducted research at Roscoff or that he visited the station. However, the author (unfortunately unknown) wrote that at the prince's request he installed oceanographic and chemical instruments to equip the table rented by the prince. The author was instructed to install these instruments "after the principles in use at the [oceanographic] museum in Monaco." Other tasks included the instruction of personnel on oceanographic procedures such as making soundings, taking water samples, taking temperature readings, collecting plankton, and analyzing the chemistry of sea water.[116] It remains difficult to assess the degree to which Albert influenced the direction of scientific work at Roscoff (or any other French marine station). He was but one of several philanthropists who supported the station; he, like Roland Bonaparte, Belgian botanist-philanthropist Jean Chalon, and members of the Rothschild family, was recognized by director Yves Delage primarily for financial contributions, not intellectual ones.[117]

A Vision of Global International Science

Prince Albert's most ambitious international project was compilation of a global bathymetric chart—the General Bathymetric Chart of the World, also known as GEBCO.[118] By 1895 sufficient sounding measurements had been made in international waters by naval vessels, telegraph companies, and oceanographic expeditions to suggest that a global map of submarine topography was feasible.[119] The main difficulty was to standardize the available sounding data. Stimulus for the project came from the Seventh International Geographical Congress in Berlin in 1899 at which a commission was established with the task of creating such a chart. Initial progress was slow until Prince Albert agreed to become chair of the commission and, crucially, to assume all financial costs. The first meeting of the commission took place in Germany in April 1903. Julien Thoulet, who had created a set of guidelines for the map in 1901, and Charles Sauerwein, a French naval officer and member of Albert's staff, assumed responsibility for the work.[120] Sauerwein oversaw the assembly of a vast amount of bathymetric data from hydrographic offices, oceanographic expedition reports, and cable-laying companies. With a team of aides he prepared preliminary drafts of the compiled data, and a first manuscript was completed in June 1904. A copy of the chart was presented to Prince Albert in May 1905. Shortly thereafter, however, a prominent French geographer, Emmanuel de Margerie, published a critique of the chart in the leading French journal of geography claiming it was flawed by inconsistencies in terminology and by transcription errors from the original sources.[121] The debacle permanently soured the relationship between Julien Thoulet and Prince Albert.[122] Yet although de Margerie's critique tarnished the project, Prince Albert did not abandon it, and subsequent versions of the bathymetric chart were greatly improved, although publication was delayed by the outbreak of World War I.[123]

Conclusion

The war horrified Prince Albert as he witnessed the disintegration of the peace he had long worked to build. Siding with the victims of German aggression, he wrote a lengthy and scathing critique of Germany in a treatise addressed to the Kaiser titled *The German War and Universal Conscience*.[124] When President Woodrow Wilson condemned Germany's policy of unrestricted submarine warfare, Prince Albert wrote to lend his support: "As a sovereign Prince, as a navigator, as a scientist, I adhere to the protest you have made with fine sentiment and human dignity against offenses committed by German arms against the rights of neutrals, the honor of mariners and public conscience."[125] During the course of the war all his oceanographic projects came to a halt.[126]

In 1921, the year before his death, Prince Albert made a final voyage to the United States to receive the Agassiz Gold Medal awarded him by the National Academy of Sciences.[127] The *Times Herald* reported that at the age of seventy-three, the "veteran prince" "fairly radiates health." Prince Albert complained to the journalist that he had not enough time "to accomplish all [he wished] to accomplish."[128] In his address to the academy the prince warned of the increasing destruction of global fisheries resulting from the use of steam trawlers.[129] He advocated for the creation of marine reserves that he hoped might be established through international agreement. At the end of his life, despite the war, Albert never truly abandoned his faith in science. As he explained in his speech,

> I have penetrated as far as I could into oceanography, where I sensed slept the solution to the big questions of biology; where I perceived the most powerful domain of the physical and chemical phenomena that gave birth to the propagation and evolution of living beings. And the more the Science on this terrain developed, the more it confirmed the formation of a new philosophy that will yield to our successors an enlarged view of the connections of the living world.

The more we compare the conditions of ocean life with those of terrestrial life, the better we sense that the principal forces of superior organisms found an initial power in the sea that was capable of furnishing the formula that progressively led to the human brain. In the future, that superiority will mount even higher as a more scientific mentality comes to possess the power of mastering human societies.[130]

When Albert died the following year, the fate of Monaco was again in jeopardy. His son Louis II inherited the throne. But Louis, estranged from his father since his parents' divorce, did not share Albert's passion for oceanography. One journalistic treatment of the succession (accompanied by a cartoon titled "The Black Wins," showing Albert slumped in his chair and death standing across from him at a gambling table) stated simply, "In contrast to his father, Prince Louis, who succeeds to the throne, . . . has no great love for the sea. He is proud of his title of colonel in the French Army and delights in military science."[131]

Perhaps foreseeing his son's disinterest in marine science, Albert had attempted to secure the future of his oceanographic program. In his will he bequeathed 700,000 francs to his longtime collaborator Jules Richard, of which 600,000 were intended to complete "the scientific and literary work" still unfinished.[132] This included the results of Albert's oceanographic cruises and an improved bathymetric chart of the oceans. After Albert's death, however, the network of collaborators on which Richard had once relied fell apart. The Great Depression further diminished the value of Albert's bequest, delaying publication of an improved bathymetric chart. New regional charts were produced abroad using techniques developed during World War II, and these charts effectively made the data compiled under Prince Albert's direction obsolete.

The Oceanographic Institute in Paris and the museum in Monaco are still in operation to this day, but their international activities were greatly diminished after Albert's death.[133] With Albert's passing, a key source of support for oceanography in France disappeared. The advantage that made Albert the most influential promoter of oceanography in the late nineteenth century—a monarch's financial and political independence—also made the long-term sustainability of his program, dependent as it was on dynastic succession, untenable. Albert was successful as visionary scientist and patron because he united "ideas," "will," and wealth. The longevity of a scientific program for oceanography, however, depended on stable networks of enduring state support and international organizations.

In the aftermath of World War I oceanographic surveys again resumed. The birth of submarine warfare stimulated military and scientific interest in the deep ocean, and technologies developed for submarine detection provided new means for mapping the deep. But the war had been a blow to internationalist aspirations in Western Europe. Instead, as we see in Chapter 3, scientists on the far side of the world took up the internationalist vision. Scientific investigation of the Pacific, they hoped, would inspire a new international community, bound by shared geography and peacefully linked by the mutually advantageous project of exploring a vast new ocean terrain.

3

Scientific Internationalism in a Pacific World

> The kinship of peoples over the seventy million square miles of Pacific seas becomes evident in the way in which they lend themselves to fusion with modern scientific progress.
>
> Hubert Work, *secretary of the Interior*
>
> The atomic bomb is a wonderful oceanographic tool.
>
> Roger Revelle

The Promise of Pacific Oceanography

In Chapter 2 I examine the growth of the marine station movement and the importance of Prince Albert I of Monaco as an early patron of international marine science—both reflections of a humanist project aimed at improving society through the advancement of science and tied to aspirations of a future free from military conflict. I also stress the increasing significance, by the early twentieth century, of international scientific networks for the coordination of large-scale scientific marine surveys. Having examined developments in Western Europe, the North Sea, and the Atlantic, we now shift focus to the Pacific. In the twentieth century, the promise of a new frontier for scientific discovery in the Pacific basin seized the imagination of scientists and helped reshape Americans' perception of their place in the world.

When the first European voyages of discovery to the Pacific region set out in the eighteenth century, their primary purpose was mapping and land-based surveys. Only toward the end of the nineteenth century did the prospect of a scientific survey of the Pacific Ocean come to be seen as both feasible and valuable. In September 1873 a US naval vessel, *Tuscarora*, set out from San Francisco Bay in hopes of ascertaining a route for a transpacific submarine telegraph cable between the United States and Japan. As an observer noted at the conclusion of the voyage, "Old Ocean, who for ages has stubbornly resisted all attempts to penetrate the secrets buried in the bosom of his waters, has been conquered, and the way is now opened by which his innermost recesses may be explored."[1] Also writing in 1873, the chief scientist of the *Challenger* expedition, Charles Wyville Thomson, observed,

> The Atlantic Ocean, with the accessible portions of the Arctic Sea, has naturally, from the relation in which it stands to the first maritime and commercial nations of the present period, been the most carefully surveyed.... We have still but scanty information about the beds of the Indian, the Antarctic, and the Pacific oceans.[2]

In the wake of successive oceanographic expeditions in the Atlantic, the absence of data from the Pacific was increasingly evident to scientists in Europe and North America. Yet even for many naturalists in the United States, the Pacific remained inaccessibly remote. Alexander Agassiz, in an 1881 letter to Thomson, wrote there was "some chance of having a government vessel off Panama" later that year; he hoped to "persuade [the] Captain to let [him] go to the Galapagos."[3] In the end, Agassiz's first oceanographic voyage to the Galápagos Islands was delayed until 1891.[4]

Despite the allure of trying to emulate the scale of the British *Challenger* expedition, launching comparable missions from Europe to the Pacific remained prohibitively costly. Steamships could travel quickly and carry out oceanographic work in the open ocean, but they were limited in their range, requiring access to coaling stations.[5] Furthermore, travel around Cape Horn was a time-consuming and notoriously risky endeavor. In 1913, however, the Pacific Ocean suddenly became a much more tempting research destination. On October 10 of that year President Woodrow Wilson sent a telegraph signal from the White House, remotely detonating eight tons of explosives in Panama, destroying the final land barrier between the Atlantic and Pacific, and completing the Panama Canal.[6]

In a May 20, 1913, letter to the British oceanographer William Herdman, the Scottish oceanographer John Murray, veteran of the *Challenger* expedition, exclaimed, "Could I afford it at present, I would be off to the Pacific in a Diesel-engined ship!!"[7] Murray had begun to plan a Pacific-bound oceanographic expedition in 1911 but encountered difficulties in raising sufficient funds. His sudden death in an automobile accident in March 1914 prevented him from realizing his Pacific ambitions.[8] But he was not the only oceanographer who had turned his attention to the Pacific.

As discussed in Chapter 2, World War I forced a suspension of most oceanographic research in Europe. When the war ended, however, the Pacific seemed to offer new terrain for discovery and, for the vanquished, opportunities to restore national prestige. The German oceanographer and director of the Berlin Institute of Oceanography, Alfred Merz, began planning an expedition for the systematic study of the Pacific basin in 1921, using the *Meteor*.[9] Merz proclaimed that the Pacific offered "an almost infinite field, which has been there all along, inviting large-scale research." A German Pacific oceanographic expedition, a "great cultural act," would both reinvigorate German science and revive nationalistic pride in the wake of military defeat.[10] Yet despite Merz's best efforts, the *Meteor* expedition was eventually forced to limit its region of study to the Atlantic, the cost of refitting the vessel for a voyage to the Pacific having proved prohibitive.[11] The *Meteor*'s confinement to the Atlantic ultimately required a restructuring of the expedition's scientific goals. Whereas in the Pacific an "open station network" might "have yielded fundamental results," the morphology and hydrology of the Atlantic basin were already well established. Owing to this wealth of existing data, Merz devised a program using a "close-knit network of stations" to measure ocean circulation using quantitative methods.[12]

An American Pacific

In the aftermath of World War I, European oceanographers imagined the Pacific as a tantalizingly remote *mare incognitum*, but Americans were turning their attention to the scientific study of the great ocean.[13] World War I stimulated the US economy, boosting the agricultural and fisheries industries. During the war, an article in the *New York Tribune* by Theodore M. Knappen reported that the California fishing industry was growing to "undreamed-of proportions." In Seattle, Knappen cheerfully reported, whale meat was being consumed at a rate of eight tons a day at a cost of ten cents a pound. Wartime demand had given "new impetus" to the tuna and sardine industries. The Pacific was being transformed into an American breadbasket. "Certain waters of the Pacific can be made to produce more food to the square mile than a section of wheat land," he wrote, adding that "here, again, the Kaiser and his war are proving to be very efficient American industrial agencies.... Germany is waging this war to fill its larder but is really spreading the American table."[14] The Pacific fisheries industry was growing exponentially and scientists working on the western coast of the United States recognized the need for comprehensive scientific study of the Pacific Ocean. Other events and actors, however, allow us to place these Pacific aspirations in a broader context.

In 1939 my grandmother, Elizabeth Haverstock, had just moved from the sleepy outskirts of Minneapolis, Minnesota, to the exciting multi-ethnic city of San Francisco. This was also the year of the Golden Gate International Exposition, a celebration of the completion of what were then the two longest bridges in the world, the San Francisco to Oakland Bay Bridge and the celebrated Golden Gate Bridge, "gateway to the Pacific." Events associated with the exposition were held all over the Bay Area. Among our family papers is a ticket for the Fantasia Pacifica pageant and ball, a "benefit for artists," held on Friday night, April 21, 1939. The ticket notifies guests that costumes were to be "restricted to the peoples and land of the Pacific Basin.... None other can be admitted to the dance floor." My grandmother must have enthusiastically taken to the floor that night because there are several photographs of her in adventurous costume, as well as a letter from her aunt that teasingly admonishes, "Shades of Salome! Would your mother do a back-flip could she have seen her only daughter clad in the seventh and last veil, a string of pearls, and a tin bra, staggering home at four a.m.!"

While youthful inhabitants of San Francisco celebrated Pacific cultures through costume pageants, Bay Area scientists were planning more serious events—also tied to the Golden Gate Exposition. On July 24, 1939, the Sixth Pacific Science Congress convened on the Berkeley campus of the University of California, and Dr. Herbert Gregory (1869–1952), director of the Bishop museum in Honolulu, delivered the opening address. As were my grandmother's, Gregory's origins were far from the Pacific. Born in Middleville, Michigan, he completed his bachelor's degree and doctorate at Yale, graduating in 1896. He stayed on at that university as an instructor of biology, physical geography, physiography, and geology, retiring in 1936. Early in his life he developed a fascination with the Pacific, and it was this passion that most shaped his career. As a colleague later recalled,

> [Gregory] used to tell of the map of the world which he and some friends drew early in the twentieth century, coloured to show the stage of knowledge reached in natural history. For the Pacific, the area coloured red to indicate exploratory work was huge; blue for reconnaissance appeared in detached patches and strips; whereas yellow, for detailed knowledge was mostly in dots. They estimated that the adequately investigated part of the Pacific was the equivalent of a plow furrow across a 20-acre lot.[15]

One of the original founders and first president of the Pacific Science Congresses, which began in Honolulu in 1920, Gregory was a principal organizer of its 1939 meeting. Although the conference was officially arranged under the auspices of the National Research Council, it was largely under the supervision of the Committee on Pacific Investigations, which Gregory chaired.[16] Gregory consistently used the Pacific congresses, whose previous meetings had been held in Honolulu, Melbourne, Tokyo, Batavia, and Vancouver, to promote collaborative scientific exploration of the region. "We are going to become Pacific-minded, instead of Atlantic-minded," Gregory had declared at the Fifth Pacific Science Congress in Vancouver in 1933. "In the generations to come this is the part of the world that is going to see the greatest progress, and I hope the greatest welfare of peoples."[17] It was in this spirit that Gregory outlined his vision for an international cooperative scientific investigation of the Pacific, focused on shared, ocean-based, geographic space. The sixth congress, as the first to be held in the continental United States, provided the organization with enhanced opportunity to receive recognition from the continental American public. After giving an overview of the history of the meetings in his opening address, Gregory assured his audience that it was "safe to say that since the meeting of the First Congress more has been learned of the structure and life of the Pacific than in all preceding time."[18]

Although Gregory might have viewed the sixth congress as marking the successes of a project begun in the early 1920s, the commercially oriented Golden Gate Exposition was, in fact, a lifeline for a project that had faltered. In the years leading up to the congress, organizers even feared that a sixth meeting would be impossible. By 1939, the political landscape of the Pacific was beginning to unravel. The fifth meeting in Vancouver had resolved that "preference be given to one of the small islands in the Pacific, or to some place on the Asiatic mainland" as a meeting site; however, early proposals to hold the meeting in Hong Kong and Fiji had to be abandoned following local political unrest. Gregory intervened in July 1938 to suggest San Francisco as the meeting location. As it turned out, funding for the congress was to be partially provided by the Golden Gate Exposition, organized by Philip N. Youtz, the exposition's director of the Department of the Pacific Area.[19]

There is good reason to invoke the image of my grandmother at a costume ball alongside that of Herbert Gregory opening the Sixth Pacific Science Congress. Together, these two images capture the strange mixture of popular culture and scientific aspiration that gave birth to the idea of a Pacific World in the first half of the twentieth century. How do the granddaughter of a Baptist minister from the Midwest and a Yale scientist, also from the Midwest, end up casting themselves as an imagined new person of the Pacific?

A core premise of this chapter is that the model of the Pacific World may be a useful framework for understanding the connection between science, commerce, and popular culture in the Pacific basin in the early twentieth century. In the United States, during the interwar period, science was promoted as a means of fostering peaceful commercial relations among nations bordering the Pacific basin. Emphasis on the interwar period represents a departure from most scholarship on twentieth-century oceanography. Although it may be true that the ocean-oriented and militarily driven science of World War II and the Cold War dwarfed the scientific accomplishments of the interwar period, prewar ocean science should not be dismissed as little more than a prelude.[20] The significance of naval power in postwar ocean science cannot be minimized, but interwar Pacific science should be examined with reference to the rhetoric that supported it. Two broad questions must be asked: What did interwar science look like? And how was it tied to the commercial and popular culture of the 1920s and 1930s?

Science and the Making of the Pacific World

Even before the mid-twentieth-century rise of big science, oceanography relied on extensive field observation and demanded collection of data on a large scale.[21] When trying to understand why scientists working in the Pacific basin sought to institutionalize international cooperative efforts it is helpful to consider what Helen Rozwadowski has termed the "environmental necessity" of conducting science at sea.[22] In her study of the formation of the International Council for the Exploration of the Seas, Rozwadowski suggests that marine scientists have long shared the conviction that oceans could be systematically studied only by international collaborative networks for the simple reason that fish and water masses know no national frontiers. In short, science at sea requires observation over a geographic space surpassing the monitoring capabilities of any single nation.[23]

In the Pacific, calls for science of an international scope were often made with appeal to national interests. From the vantage point of the United States, an orientation toward the Pacific can be traced back to the early 1800s, when national interest was routinely identified with commercial advantage and scientific progress. Thomas Jefferson's orders to Meriwether Lewis in 1804 commanded that his expedition find the river "offering the best communication with the Pacific Ocean" for "the purposes of commerce."[24] In the 1840s, Charles Wilkes's South Seas Exploring Expedition sought to "extend the empire of commerce and science" into the Pacific while "carrying the moral influence of our country wherever our flag has waved."[25] During the 1840s, the celebrated pioneer of marine science Matthew Fontaine Maury published a series of editorials advocating naval reform, in which he decried the lack of cartographic data dealing with America's western oceanic frontier:

If you have a map of the world at hand turn to it, and placing your finger at the mouth of the Columbia River, consider its geographical position and the commercial advantages which, at some day not far distant, that point will possess. To the South, in one unbroken line, lie seven thousand miles of coast indented with the rich markets of Spanish America—to the West, Asiatic Russia and China are close at hand—between the South and the West, are New Holland and Polynesia; and within good marketable distance are all the groups and clusters of islands that stud the ocean, from Cape Horn to the Cape of Good hope, from Asia to America. Picture to yourself civilization striding the Rocky Mountains, and smiling down upon the vast and fruitful regions beyond, and calculate, if you can, the importance and future greatness of that point to a commercial and enterprising people. Yet the first line in the hydrography of such a point remains to be run.[26]

Scientific progress was tied to the expansion of American empire to the Pacific basin, and scientific knowledge was cast as a gift of the United States to other Pacific nations. When Commodore Matthew Perry signed a trade agreement with the emperor of Japan in 1854, his diplomatic gifts to the emperor included a short, circular railway track and steam engine, an electric telegraph, a printing press, a set of John James Audubon's American ornithology books, plates of American Indians, maps, and agricultural implements with "all the modern improvements." A writer for the *New York Times* noted that although the Japanese were initially "astonished" by the telegraph, "they will speedily understand it, and may possibly by this time be laying down the wires for themselves."[27]

By the early twentieth century, a new perspective on science in the Pacific was emerging. No longer tied solely to the expansion of American military power, this view of Pacific science embraced an ideal of peaceful international collaboration. Scientists, business leaders, and politicians imagined science as the linchpin in the formation of a new transnational community, based on a shared geographic location and linked by common need to manage and exploit the resources of the Pacific marine environment. As set out in the resolutions of the Fourth Pacific Science Congress held in Indonesia in 1929,

Whereas the oceanographic problems of the Pacific Ocean because of its vastness cannot be solved in any reasonable time by any one institution or any one country and in view of the efforts that are being made by the International Committee on the Oceanography of the Pacific to bring about co-ordinate action by the different countries bordering the Pacific and by other countries interested in the necessity for the complete exploration of that Ocean, it is proposed: That every endeavour should be made in the direction of urging the need for the setting up of oceanographical stations in the various lands bordering on or lying within the Pacific Ocean where such stations do not already exist, with a view to ensuring a complete international oceanographic survey of the Pacific Ocean.[28]

At the 1933 meeting, the American oceanographer Thomas Wayland Vaughan (1870–1952) was happy to report that there were now nationally run marine stations throughout the Pacific basin, in New Zealand, New Caledonia, Australia, Indonesia, French Indochina, the Philippines, China, Japan, Siberia, Canada, and the United States. He suggested that a new "international station" be constructed on "some easily accessible island in the tropical Pacific."[29]

The 1933 proceedings reports on oceanographic work carried out by the United States were disproportionally lengthy in comparison to other national reports, something Vaughan noted with regret. In apology, he argued that their length was due to the facility with which American institutions were able to exchange reports of their activities.[30] In establishing marine stations, too, the United States had been more active than other nations. With an institutional network in place by the early 1930s, all marine scientists working in the Pacific basin were eager to form international partnerships.[31]

A Letter from Japan

The archives of the Scripps Institution of Oceanography contain a 1921 letter from the Japanese marine scientist Kamakichi Kishinouye, professor of fisheries at the Imperial University in Tokyo, to the Scripps oceanographer George McEwen. In it Kishinouye provides a short account of "oceanographical work in relation to fisheries" in Japan. Little is available in English about the history of marine science in early twentieth-century Japan. Kishinouye's letter sheds light on a little-known chapter in the history of Pacific oceanography and provides an early example of scientific cooperation between Japan and the United States. Kishinouye writes about Japanese marine investigations,

> The oceanographical work in relation to fisheries was done at first in 1892 to know the direction of ocean currents round our islands by drift bottles. In 1900, observations of the temperature, salinity, meteorological data, plankton, etc. were undertaken at five stations ... four times a year. ... Besides this several trips of a steamboat were tried in the sea near the Tokyo Bay to carry on intensive oceanographical investigation. In 1918, a special steamboat was constructed for ... investigations, and since 1919 she is engaged in this work. Thus so far we are observing our coastal waters, ocean-currents near our coasts, and the plankton in these waters; but the observation and the study of the high seas are very scarce. Therefore we hope that someone would undertake expeditions to explore such region minutely.[32]

Kishinouye and McEwen were both prominent scientists. Kishinouye's obituary in *Science,* published February 14, 1930, describes him as one of the "leading scientific men of Japan," an expert on coral and mackerel, remembered as "a good example of the courteous Japanese gentleman of the old school." He died in November 1929 while on a research trip to China.

McEwen, who had come to Scripps as a graduate student in 1908, was a leading figure in the early development of that institution. His influence was crucial in transforming what had been a small seaside laboratory into the vanguard center for scientific investigation of the Pacific. The historian Eric Mills writes that no other North American oceanographer "could match the combination of ability and opportunity that met in McEwen, allowing him to apply mathematical physical oceanography to North American waters."[33] McEwen was a vocal proponent of international cooperation, pointing out that a successful model for international cooperation in the Pacific had already been established for the Atlantic basin in the form of the International Council for the Exploration of the Seas, founded in 1902. In a paper delivered at the 1919 meeting of the Pacific Division of the American Association for the Advancement of Science, McEwen lamented that no comparable organization existed for the Pacific: "For every paper pertaining to the Pacific, there are scores of voluminous reports devoted to the North Atlantic and neighboring seas and gulfs."[34] McEwen was primarily a physical oceanographer, but fisheries research ultimately helped form a collaborative program in the Pacific.[35]

That a Japanese marine scientist should have been in correspondence with an American marine scientist at Scripps in the early 1920s is an important reminder of the efforts made in the early twentieth century to build a program for international scientific cooperation in the Pacific prior to World War II. Some Japanese scientists studied at Woods Hole and Scripps in the United States, and American scientists visited marine institutions in Japan. Scripps founder William Emerson Ritter (1856–1944), George McEwen's mentor and another proponent of international cooperation, visited the University of Tokyo's marine biological laboratory at Misaki in 1906.[36] In the 1920s and 1930s considerable headway was made toward a cooperative program for oceanography; the Pacific Science Association was created in 1926, and a meeting of the Pan-Pacific Science Congress was held in Tokyo later that year. At the Tokyo meeting, the oceanography sessions were overwhelmingly composed of Japanese and American scientists.[37]

For a brief period at the beginning of the twentieth century it appeared that a new program for collaborative Pacific research was on the verge of exponential growth. "Oceanography is the science that coordinates the results of research done in all branches of science, as these pertain to the ocean, its contents and its boundaries" declared Charles McLean Fraser, the head of the department of zoology at the University of British Columbia, in an article published in *Scientific Monthly* in 1937.[38] Whereas in the past scientists had been able to collect only "scraps of information," Fraser explained—the scientific disciplines metaphorically confined as in "water-tight compartments"—this had come to an end. Now, Fraser declared, "the most striking single feature in the progress of science . . . [is] the spirit of cooperation and coordination." Oceanography was "a product of the present century; Pacific oceanography, largely that of the last two decades." In sum, "there was no real cooperation in Pacific work before the First Pacific Science Congress in Honolulu in 1920." The birth of Pacific oceanography could be traced to 1920, but Fraser insisted that "more had been accomplished in Pacific oceanography during the preceding four years, 1929–1933, than in all previous time."[39]

Reframing Science for the Pacific World

The dream of a unified Pacific World, and even of pan-Pacific citizenship, was first conceived by civic boosters, ethnic leaders, and journalists of the region in the early twentieth century.[40] This coincided with a period in which science was moving to center stage in social and political discourse. Thus, for much of its history, the Pacific World has been partly viewed through the lens of science. Most revealing on this point is a speech delivered by the president of the newly minted Pacific Scientific Institution (based in Honolulu), William Alanson Bryan, at a meeting of the American Association for the Advancement of Science in Chicago in 1907, in which he addressed the urgent need to advance the study of the Pacific in its entirety. He argued that, so interconnected were the objects of study in the region, even an ethnological survey required information that was still lacking on the "oceanology, climatology, geology, zoology, and botany of the entire ocean."[41]

In his speech, Bryan went on to discuss his plans for a scientific expedition on a yacht of "five to seven hundred tons capacity," promising that "the study of each island will be made with an understanding of the great ultimate object, namely, knowledge of the Pacific Ocean as a whole."[42] In his view, such knowledge could be gathered only as part of a transdisciplinary project.[43] Western science, relocated to the Pacific, would take on new configurations.[44] We find evidence of this transformation in world's fairs on the West Coast of the United States, where organizers presented scientific work in the Pacific region as a transnational enterprise and as key to forging a pan-Pacific community.

To mark the opening of the Panama Canal in August 1914, a celebratory exposition was held in San Francisco the following winter. In conjunction with the Panama-Pacific International Exposition, a scientific conference was organized. It was the first meeting of the American

Association for the Advancement of Science to be held on the West Coast. In an address delivered at that meeting, "Problems of the Pacific Islands," the Harvard professor of geology Reginald Aldworth Daly presented a proposal for approaching "the last great field for scientific conquest."[45] "Is not the piercing of the Panama a suggestion, a brilliant symbol, for American Geographical science?" he asked his audience. He suggested the Pacific be divided into separate regions that would fall under the responsibility of different nations to investigate.

> To the government bureaus, or scientific bodies of the different nations concerned, may be assigned the duty of scientific exploration in the Aleutian, Kurile, Japanese, and East Indian archipelagoes; but Polynesia, Micronesia, and Melanesia need different treatment ... the thorough study of these oceanic islands offers a highly desirable program for large-scale private enterprise in science, and ... is highly appropriate for American enterprise in particular.[46]

At this meeting, Pacific marine resource conservation was also a matter of foremost concern. Presenters offered data showing that both the Alaskan seal hunt and the Pacific crab fishery were showing signs of overexploitation. As Walter P. Taylor, curator of mammals at the University of California, Berkeley, noted, "It is coming to be realized that, particularly in a democracy, a special obligation to furnish leadership in movements for the perpetuation of the native fauna rests upon the professional zoologist."[47] While scientists presented their papers at the conference, public visitors touring the fanciful concession grounds of the Panama-Pacific Exposition were treated to a different, more bountiful, vision. One of the popular attractions, described by newspapers as "conceived for amusement and instruction," was a ride called the Submarines (see Figure 3.1). It transported visitors on a simulated undersea journey, "around the world, beneath the waters of the seven seas, going through a succession of winding subterranean caverns and seeing many unheard of wonders of the mighty deep."[48]

The themes introduced in the Panama-Pacific Exposition were repeated in the Golden Gate Exposition in 1939. Robert Rydell, a historian of world's fairs, describes that exposition as "an imperial dream city" animated by

Figure 3.1 Entrance to the Submarines at the Panama-Pacific Exposition. Robert A. Reid, *The Red Book of Views of the Panama-Pacific International Exposition, San Francisco 1915* (San Francisco: Panama-Pacific International Exposition, 1915), 85.

"dreams of Empire."[49] He points out that many of the same organizers designed both fairs. Combining patriotic display with carnival forms of entertainment, the 1939 fair was presented as a pageant of the Pacific. It also celebrated San Francisco's claim to be a gateway to that world. In similar fashion, the San Francisco meeting of the Pacific Science Congress was a celebration of American aspirations for leadership in transnational Pacific science.

Embodying an Ocean in Pacific House

A colossal statue named *Pacifica* overlooked the Court of the Seven Seas, and at the center of the Golden Gate Exposition, on Treasure Island in San Francisco Bay, stood Pacific House, the officially designated theme building (see Figure 3.2). An excellent description of Pacific House comes from an article published in the *San Francisco Chronicle*, titled "Pacific House Popular," that is worth quoting at some length:

Figure 3.2 Statue of *Pacifica* at the Golden Gate International Exposition. Postcard postmarked from the fairground on July 2, 1939. Reproduced from the author's collection.

Treasure Island's monument to peace, ... Pacific House is the theme building of the Fair, uniting in ... brotherhood all nations served by the boundless Pacific. Thousands ... visited the glistening temple, erected in the hope of promoting understanding among races of the Pacific.... Outside the crowds may roar ... but the gorgeous Pacific parlor is somehow aloof, remote, reflective.... There is a library of rare books, a silent reading room, attentive librarians. Upstairs there is a conference room, with motion picture equipment for future illustrated lectures. But most ... will be drawn immediately to the Sotomayer fountain, ... an enormous terracotta map depicting the physical appearance of the Pacific area.... [It gives a] bird's eye view ... Every tiny island of the Pacific is shown, as well as every vast country. The ocean itself is caught in seven shades of blue showing the various depths and currents.... Trade routes ... are pictured in a stained glass window that spans the north portal of the building, its florescent colors gleaming like jewels.... But dominating the entire building are the Covarrubias murals, brilliantly colored, that picture the fauna, flora, peoples, economy and arts of the Pacific area. Directors of the building have sought to bring ... the verdant fascination of each section of the Pacific by installation of plants indigenous to North and Central America; South America, Australia, New Zealand and vicinities and Asia. Such is the exposition's temple of peace—dignified, elegant, possessed of an almost ineffable serenity.[50]

Pacific House offered a human-scale representation of what was in reality an enormous geographic area. Visitors could view the Pacific in miniature from a bird's-eye view and contemplate the urgent need for international peace and cooperation in a bounded region. As the exhibits made clear, the Pacific was a vast interconnected system. Science, the project of mapping, classifying, and studying these connections, was presented as the handmaiden of transpacific communication, cooperation, and peace.

Pacific House was the main venue for public lectures organized by the Pacific Science Congress. Topics included "Public Health Conditions in China," "Origins of Cultivated Plants in Relation to the Origins of Civilization," "Explorations in the Lesser Sunda Islands," "The Structure of the Pacific Ocean as Indicated by Earthquakes," and "Ocean Currents of the Pacific and Their Bearing on the Climates of the Coasts," this last delivered by the Scripps oceanographer Harald Sverdrup (1888–1957).

All events in Pacific House took place under the direction of Philip N. Youtz, a man of eclectic experience. After returning to the United States following several years in China, Youtz served as curator at the Pennsylvania Museum of Art and then assistant director of the Brooklyn Museum of Art before becoming director of Pacific House in 1938. As the fair was winding down, Youtz drafted "Proposal for a Permanent Pacific House," arguing that America's "manifest destiny" lay in the extension of a "Good Neighbor policy through the Pacific Hemisphere":

> The institution might be the permanent home of certain scientific organizations with a secretary to keep in touch with the research work carried on in each of the Pacific countries and to correspond with the different scholars.... There would be an unusual opportunity for publication of the contributions of those men in more than one language ... [and] Pacific House might well sponsor scientific expeditions to different parts of the area and serve as a clearing house for the findings of these investigations.[51]

San Francisco, Youtz argued, was "in a position to become the capital of a new Pacific empire."[52] But none of this was to be. On September 29, 1940, the fair closed. The navy quickly moved to acquire Treasure Island. On December 7, 1941, Japan attacked Pearl Harbor and the United States entered World War II. The fair buildings on Treasure Island were dismantled, replaced by a naval base, and all planning for future Pacific Science Congress meetings ceased. The dream of a Pacific World defined by peace, scientific cooperation, and mutually beneficial commerce had been supplanted by a nightmare: a Pacific theater of war. President Franklin Delano Roosevelt, in his fireside chat of February 23, 1942, asked his listeners "to take out and spread before [them] a map of the whole earth," informing Americans that the oceans had now become "endless battlefields on which we are constantly being challenged by our enemies."[53]

The Lights Go Out on Treasure Island

At the exposition's closing ceremonies on October 29, 1939, Leland Cutler, president of the exposition, remarked that the fair had been "the dream of many."[54] The war not only ended the dream of a peaceful Pacific but also reshaped memories of what the fair had been. As the military author of an official naval history of Treasure Island recalled in 1946, "Yes, Treasure Island had its days of festivals—but they were ominous days, and there was treachery at the feast. Supposedly fostering better relations with the nations bordering on the Pacific Ocean—the South Seas, the Antipodes, Central and South America, and the Orient ... there was one amongst them whom we knew not."[55] Some fair visitors remembered with new suspicion the presence of Japanese tourists taking photographs of the bay while visiting the fairground.[56] For others, the closing of the fair simply marked the end of a peaceful era. As one visitor later recalled, "I remember going with my husband up to Telegraph Hill to watch the lights go out on Treasure Island. Soon after, we would go to the same spot to see ships depart for the South Pacific."[57]

Shortly before the end of the war, San Francisco hosted the first meeting of the United Nations. Secretary of State Edward Stettinius proposed San Francisco as the site of the meeting, inspired by a dream. "I saw the golden sunshine," he recalled; "I could almost feel the fresh and invigorating air from the Pacific."[58] The first meeting went ahead as planned on April 25, 1945, but under the watchful gaze of antiaircraft gunners on the lookout for Japanese incendiary balloons drifting over the ocean on the eastward breeze.[59]

Only in March 1949 did the Pacific Science Congress finally reconvene, this time in Auckland and Christchurch, New Zealand. Herbert Gregory again delivered the opening address and once more stressed the importance of scientific cooperation, but this time some representatives were noticeably absent. There were no delegates from Japan, the host of the Third Pacific Science Congress, in 1926. Following Gregory's speech, the president of the congress, R. A. Falla, director of the Dominion Museum of Wellington, New Zealand, took the stage:

> The coordinating responsibility of the Pacific Science Congress had grown immeasurably in the interval between the sixth and seventh.

Not only was the field of science itself extending further into the knowledge and use of ionosphere and atomic structure alike, but once again war and its aftermath had stimulated the organization and direction of research to a degree that all our planning of the past could hardly have foreseen. One could detect in the program now before members an increased proportion of papers dealing with wide application of the physical sciences. Other cooperative sciences, like oceanography, would appear also to be past the pioneering stage in which, as far as the Pacific was concerned, they were in 1939.[60]

War had reshaped the nature of science in the Pacific basin, reoriented the scientific landscape for what would soon become the Cold War. Pacific science no longer depended on "large-scale private enterprise" but on military patronage. The utility of oceanographic science for naval warfare had been proved in the Pacific theater.[61] In the wake of World War II, the United States and the Soviet Union dominated exploration of the marine environment. Perhaps the most enduring reminder of the lost aspirations of Pacific scientists during the interwar period is a note that to this day is displayed in many American marine stations. This note, written by the Japanese marine biologist Katsuma Dan, Allied troops found posted to the door of the Misaki Marine Biological station in Japan when they arrived on September 2 1945, the day of Japan's surrender:[62]

> This is a marine biological station with her history of over sixty years. If you are from the Eastern coast, some of you might know Woods Hole or Mt. Desert or Tortugas. If you are from the West Coast you may know Pacific Grove or Puget Sound Biological Station. This is a place like one of these. Take care of this place and protect the possibility for the continuation of our peaceful research.

Time magazine republished the note on December 10, 1945, in "Appeal to the Goths." The military officer who retrieved it sent it on to the Marine Biological Laboratory in Woods Hole, Massachusetts, where it was copied and distributed to other marine stations around the country. To this day, the original still hangs on the wall outside the library of the Woods Hole Marine Biological Laboratory.

The Fair as Research Center

Although identification of the Pacific as terrain for scientific internationalism waned as a result of the war, the idea of the fairground as space in which scientific data could be aggregated and distributed survived. In fact, we see the legacy of the conflation of fair and research center when we examine the science center movement in the second half of the twentieth century. The two best examples of science centers on the West Coast of the United States are the Exploratorium in San Francisco and the Pacific Science Center in Seattle. Both were built in the postwar, post-Disneyland era of the early 1960s.[63] The 1962 Seattle World's Fair, called the Century 21 fair, provided the seedbed for the Pacific Science Center, still in operation today. The initial conception of what a science center could be was very different from the edutainment aimed at children now associated with such institutions.

The victorious emergence of the United States from World War II bolstered public confidence in American science and industry. The optimism of American science as an "endless frontier" was famously articulated in Vannevar Bush's call in 1945 for increased government support.[64] Bush predicted, "Advances in science when put to practical use mean more jobs, higher wages, shorter hours, more abundant crops, more leisure for recreation, for study, for learning how to live without the deadening drudgery which has been the burden of the common man for ages past."[65] Yet the threat of conflict with the Soviet Union and advances in atomic science also fostered fears of nuclear apocalypse. The popular perception of science in the early 1960s was thus informed by diametrically opposed visions of how science could shape the future.

As the urban historian John Findlay has observed, Cold War tensions transformed the Seattle fair "from a 'Festival of the West' to 'America's Space Age World's Fair.'"[66] Initially conceived as a way for local businesses to generate tourism revenue, the fair soon became a platform for the federal government to promote the achievements of federally supported research programs.[67] President John F. Kennedy declined to attend the opening ceremony, claiming a "cold," while in reality dealing with the Cuban Missile Crisis. In his place Vice President Lyndon B. Johnson delivered an address at the dedication of the fair's NASA Space Exhibit on May 10, 1962. "We are hopeful of achieving fruitful cooperation with the Soviet Union," he told the assembled crowd and announced the possibility for joint research in "communications, weather forecasting, mapping the earth's magnetic fields and space medicine."[68] Others, such as Secretary of State Dean Rusk, used the fair as an occasion for painting a more dystopian view. Without "international law and supervision" Rusk warned, "the frontiers of space might be pierced by huge nuclear-propelled dreadnaughts, armed with thermonuclear weapons. The moon might be turned into a military base. Ways might be found to cascade radioactive waves upon an enemy. Weather control might become a military weapon."[69]

The "Futurist": Athelstan Spilhaus

A centerpiece of the fairgrounds was the Federal Science Exhibit. Designed by Seattle-born Japanese American architect Minoru Yamasaki, it consisted of five buildings grouped around a series of pools, fountains, and gothic arches.[70] The person in charge of the exhibit was Athelstan Spilhaus. After obtaining a master of science in aeronautical engineering from MIT in 1933, Spilhaus returned to his native South Africa, working for its military on the meteorology of the upper atmosphere. In 1936 he returned to the United States as a research assistant at the Woods Hole Oceanographic Institution. There he developed oceanographic instruments, most notably the bathythermograph, used to measure variation in water temperature with depth. Spilhaus's device had important naval applications during World War II, providing crucial calibration information for the sonar needed to locate enemy submarines. During the war, Spilhaus served in the US Army, teaching meteorology and traveling throughout Europe and China, eventually achieving the rank of lieutenant colonel. After the war, Spilhaus obtained a doctorate in oceanography from the University of Cape Town before returning to the United States, where he founded and chaired a department of meteorology and oceanography at New York University.[71] He eventually left NYU to become dean of the Institute of Technology at the University of Minnesota.[72]

Spilhaus described himself as a "futurist," and some of his ideas now sound quixotically utopian, such as his prediction in a 1964 issue of the *Bulletin of the Atomic Scientists* that soon "people will drive down to underwater resorts, park their submobiles, check into submarines, and participate in one of the many available recreations." In the same article he described dolphins trained to service fisheries: "One day sea 'shepherds' or 'cowboys' may ride bucking one-man submarines."[73] For over a decade, he also produced a Sunday morning comic strip, *Our New Age*, that described ongoing scientific projects and prophesied science of the future. In the early 1970s he attempted without success to build an experimental enclosed city under a massive geodesic dome in northern Minnesota.[74] Such was the man whose vision shaped the futurist science exhibits for Century 21.

A "Jewel Box" of Wonders

In an opinion piece in the *Seattle Post-Intelligencer* in advance of the opening of the fair in March 1961, Warren Magnuson, senator from Washington State, enthused that the Federal Science Exhibit "was not a museum. It was a 'jewel box' arrangement where man could look ahead to the year 2000."[75] Visitors began their tour of the exhibit with a thirteen-minute film titled "The House of Science." Spilhaus praised the filmmaker for having "talked with hundreds of scientists in many countries, and visited more than 50 laboratories with his camera."[76] Projected as a collage of images, the short film showed scientists at work in various types of settings. "A laboratory can be many things and many places," explained the narrator. As scenes of radio telescopes, rockets, space phenomena, and scientific instruments flashed by on the screen, the narrative presented an optimistic view of free and collaborative research. "Science is essentially an artistic or philosophical enterprise carried on for its own sake," the narrator explained. "In this it is more akin to play than to work."[77]

The visitors' tour proceeded into building II, History of Science, where exhibits told the "story of the development of science from the simplest beginnings to its present-day refinements." In building III, Spacearium, sponsored by Boeing, they were shown another thirteen-minute film, "Journey to the Stars," on what was then the largest screen in the world. In Building IV, Methods of Science, the largest of the exhibition halls, displays were grouped into six major subject areas: the nature of our surroundings, the sources of energy, the structure of matter, the nature of life, the functioning of living organisms, and the sources of animal and human behavior. The highlight of these displays was the "transit tracking station," funded by the Bureau of Naval Weapons and staffed by the Johns Hopkins University Applied Physics Laboratory. Visitors stood beneath a model of the Transit 4A satellite while a presenter explained how satellites allow scientists to "determine the true shape of the earth." Other exhibits detailed rocket and satellite research programs and included a diagrammatic exhibition on Project Argus, a "large-scale scientific experiment involving the explosion of a nuclear device in the upper atmosphere." This "experiment," visitors were informed, had given scientists a better understanding of the aurora.

Although funded by the Bureau of Naval Weapons, exhibits downplayed the military applications of this research. In building IV female-only "Science Demonstrators" staffed the exhibits and gave public demonstrations. Spilhaus claimed that the women "were hired for a serious purpose" and that some became efficient lab technicians.[78] In an exhibit depicting "the modern laboratory," they performed experiments on the optical nervous system of horseshoe crabs (long a favored experimental organism in East Coast marine stations but not native to the western Pacific), demonstrated the effects of radiation on the growth of mold, and showed chemical extraction techniques. Additional exhibits included a lecture hall, and a "Junior Laboratory" where hands-on exhibits were designed for children.[79]

Finally, in building V, "The Horizons of Science," visitors were conveyed on a moving carpet past exhibits, models, and films showing scenes from nature while a recorded voice asserted a linkage between science and peace. The final message seen by visitors as they left the science pavilion explained that it was "the benefits of science that men quarrel about, not science itself."[80] The same message could be found in other displays throughout the fairground. The British pavilion, for instance, showcased a hovercraft and freeze-dried food, and in the French exhibit elegant "young hostesses" led tours through an exhibit that compared the threats of "street mobs and rock-'n'-roll singers, racing cars and shouting demagogues" with "the rewards of research," exemplified by a model of Jacques Cousteau's diving-saucer submersible. In a similar gesture toward the future of marine exploration technology, the Canadian exhibit showcased the potential colonization of the Arctic with a diorama depicting "submarines shuttling beneath the ice," carrying people and supplies to dome-covered towns on a frozen sea.[81]

The Fair Closes, the Science Center Opens

Echoing the aspirations of the organizers of Pacific House at the Golden Gate International Exposition two decades earlier, the organizers of the Federal Science Exhibit discussed the possibility of turning it into a permanent institution after the Century 21 fair closed in October 1962. The proposals of the scientific advisory committee weighed the possibility of turning the science exhibits into a "Smithsonian of the Northwest," a center for television broadcasting, or a site for future science conferences.[82] Unlike Pacific House, the Federal Science Exhibit outlasted the fair to become the Pacific Science Center.[83] The most important figure in bringing this about was the marine biologist Dixy Lee Ray, who had previously served with Spilhaus on the National Academy of Sciences Committee on Oceanography.

Upon becoming director of the Pacific Science Center in 1963, Ray, already a well-known science popularizer, argued that the new institution could be "a great experiment" designed to "amuse, beguile, stimulate, inspire, [and] inform," making the people "of the Pacific Northwest scientifically the most literate" of the nation.[84] But although the future of the Pacific Science Center was secured, upon losing its connection to the world's fair it became a local institution. The internationalist rhetoric of the fairground disappeared, as did the prewar dream of a Pacific-wide identity forged through collaborative science.[85]

Conclusion

The marine sciences that emerged after World War II were shaped by the requirements of naval warfare. Marine geology, geophysics, and marine acoustics were growth fields in the postwar period. The advent of nuclear-powered submarines was accompanied by increased need for underwater maps and sonar detection. The Scripps Institution of Oceanography, in particular, benefited from military patronage. The war had demonstrated the utility of Pacific oceanographic research for naval applications.[86] Upon the war's conclusion, the remote Pacific quickly became an atomic testing ground, and its native inhabitants involuntary test subjects.[87]

A newly created Office of Naval Research began its first contract with Scripps on July 1, 1946, and soon emerged as the leading source of funding for oceanography in the United States.[88] Military and scientific interests became increasingly entwined, so that the Scripps oceanographer Roger Revelle (1909–1991) famously—if facetiously—lauded the atomic bomb "a wonderful oceanographic tool."[89] Although this new partnership required that some scientific research remain classified, the benefits of collaboration were seen to outweigh the hindrances of maintaining military secrecy. As the Scripps physical oceanographer Walter Munk later recalled, there were disagreements over declassification issues, but "this was not a battle of the Navy against the oceanographic community.... Decisions were made by an *interacting* group of Navy officers, some with deep scientific interests, and academic oceanographers who were unusually well informed on, and cared about, Navy matters."[90] With growing Cold War tensions, and ensuing military patronage, the 1950s and 1960s was a period of dramatic growth for marine science. Public interest in scientific exploration was tinged by Cold War narratives of competitive nationalism. As we see in Chapter 4, enthusiasm for the conquest of oceanic "inner space" mirrored new obsession with a race to conquer "outer space."[91]

Nevertheless, as Jacob Hamblin has argued, if "oceanography was a Cold War science ... its most crucial component was international cooperation."[92] According to him, "Oceanography presented a paradox: it was both a major recipient of military research funds and a model field of inquiry in which all nations could cooperate."[93] Yet as we have seen, internationalist oceanography predated the Cold War and reemerged after a period of wartime dormancy. From this perspective, efforts to achieve international cooperation in oceanography during the Cold War appear less as a "paradox" and more as a continuation of an enduring characteristic of marine science.

As in the period following World War I, marine science gathered momentum after the end of World War II. The "ocean frontier" became a focal point in scientific, political, and popular discourse.[94] In his State of the Union address in 1961, President Kennedy, promising improvements to national education, confessed, "We have neglected oceanography."[95] Yet as evidenced two years earlier by an article in a July 1959 issue of *Time* magazine, oceanography was hardly being neglected:

> With the ocean now transformed from barrier to a new and menacing frontier from which guided missiles could be launched upon U. S. cities, the Navy's concern with oceanography has expanded. That concern has brought U. S. oceanographers money, men and resources they never dreamed of before the war [and] made their specialty perhaps the fastest-growing science in the world.[96]

As long as the military was willing to foot the hefty bill, oceanographers accepted the funding. As Columbus Iselin, wartime director of the Woods Hole Oceanographic Institution, informed a *Time* magazine reporter, "The Cold War and the scientific effort run parallel much of the time. They're both geared toward our learning more. Each has a different motivation. One is survival, and the other is curiosity."[97] Military officials and marine scientists had a common goal: increasing the scope and scale of observation and experimentation at sea. As Robert W. Morse, assistant secretary of the navy for research and development, explained in a 1966 speech, "The scientist must understand that the Navy's problems with the environment are as large as the oceans themselves and as long range as science itself."[98] With a new military and scientific partnership securely established, a "golden age" of global oceanography had begun.[99]

Internationalism remained an important feature of the marine sciences in the postwar period, but military patronage wrought dramatic changes. The most important of these was a shift in scale of the field of observation. The Pacific became too small an area of study for an emerging superpower concerned with maintaining naval supremacy. The need for oceanographic observations on a global scale was further confirmed by several important scientific discoveries: mounting evidence in support of the theory of plate tectonics and improved understanding of ocean circulation and marine meteorology—discoveries made possible by financial and technological resources provided by the military.[100] In a sense, the era of Pacific-World science had come to a close. Despite the national interests served by military backing, the quest for an ever-expanding scale of observation ensured that scientists' aspirations for transnational cooperation would persist.[101]

Yet with the entrance of the federal government—and military—as primary patrons of oceanographic work, international collaboration was subject to new limitations. As set out in the 1966 Marine Sciences Act, collaboration "with other nations and groups of nations and international organizations" was justified "when such cooperation is in the national interest."[102] Science in the Pacific World, no longer rhetorically linked to a shared interest in securing world peace, became currency for garnering allies in an extensive geopolitical conflict zone.[103] As we see in Chapter 4, anxiety that ocean frontiers might become the decisive military and economic assets of the future profoundly reshaped imagination of marine spaces.

4

Cold War Science on the Seafloor

> Soviet Russia aspires to command the oceans and has
> mapped a shrewdly conceived plan, using science as a
> weapon to win her that supremacy.
> Warren G. Magnuson, Washington State Senator

> When divers gazed into the ocean, they saw, and probably
> continue to see, in part a reflection of their imagination.
> Helen Rozwadowski, "From Danger Zone to World of Wonder"

This chapter examines developments in marine exploration during the Cold War. I foreground the invention of scuba (self-contained underwater breathing apparatus) technology and the development of man-in-the-sea projects aimed at extending the time divers could live and work underwater. As a case study, I discuss a scientific research program called Project Sea Use (Seamount Exploration and Undersea Scientific Expedition), a collaborative effort involving private industry, state government, and military support, carried out on Cobb Seamount off the coast of Washington State from 1960 until 1975. In surveying both what was accomplished and planned projects that failed to materialize, this chapter describes the projection of future uses of the seafloor that emerged during the Cold War. Anxieties, aspirations, and intents, shared by engineers, scientists, policy makers, private companies, journalists, and science fiction writers, arose as new technologies promised unprecedented access to marine mineral riches.[1] Seamounts, in particular, possess characteristics that appealed to the period's imagination of potential scientific and geopolitical uses of the seafloor. Other environmental constraints, however, hindered the realization of this vision. The story of Project Sea Use draws attention to the multiplicity of aims driving marine research at the height of the Cold War, as well as to the environmental factors that enabled or restricted these goals.

Despite the geopolitical rivalry between the United States and the Soviet Union, international scientific cooperation in marine science was again on the rise in the 1960s, culminating in 1968 with President Johnson's call for an "International Decade of Ocean Exploration"—anticipated as an "unprecedented and historic adventure."[2] Yet although engineers produced numerous undersea habitat feasibility studies throughout the late 1960s and early 1970s, few of their proposals were implemented.[3] The failure of most of these projects is attributable to lack of funding and to little promise of financial return. Also of critical importance was that the Soviets showed no sign of reciprocating a race to occupy the seabed—although their submarines posed a challenge to American dominance on the high seas. In the absence of a government-backed "wet NASA," driven to colonize the seabed in an accelerated arms race with Russia, it fell to private corporations like General Electric, Chrysler, and Honeywell to spearhead undersea colonization. Eventually, as undersea habitats were deemed extraneous to the Cold War military effort and the promise of lucrative federal contracts faded, private investment disappeared as well.

Examining the height of the habitat proposal period is instructive. It provides context for understanding how federal, regional, and private interests cooperated in marine science and exploration during the Cold War and reveals how international policy shaped the direction of scientific research and public investment at the state level. Project Sea Use never resulted in the undersea research habitat its promoters hoped for; instead, it became a multiyear research program supported by the Washington State legislature, the navy, private companies, and a team of divers invested in the study of saturation diving and underwater medicine.[4] To explain this mobilization of workers and regional resources, we must first understand how the dream of underwater colonization became rooted in the popular imagination.

Dramatic transformations in the marine sciences occurred as a result of the wartime partnership of scientists and the navy during the 1940s. Like the physicists of the Manhattan Project, oceanographers enjoyed new visibility and prestige for the role oceanography had played in the US victory. The Pacific—a territory now firmly under the jurisdictional control of the United States—promised a new frontier for American exploration and discovery. And American marine science grew rapidly thanks to instruments and research foci developed during the war.

Prior to World War II, oceanography had been a disjointed, underfunded discipline practiced by a few visiting scientists at remote marine laboratories.[5] One of the most prominent figures of the generation of oceanographers to emerge from the war was the director of Southern California's Scripps Institution of Oceanography, Roger Revelle. As a naval lieutenant commander, Revelle had facilitated collaboration between the Navy Radio and Sound Laboratory and Scripps in studying underwater acoustics, submarine detection methods, wave forecasting, beach landings, and harbor defense systems. After the war he convinced the navy to continue to fund bathythermograph experiments, wave and current research, and sea bottom studies.

In the immediate postwar period, several undergraduate training programs in oceanography were established by American universities.[6] Marine scientists benefited from continued support of the Office of Naval Research and from the return to civilian research institutions of naval oceanographers armed with a newly legitimate voice in national political and military debates.[7] With enhanced authority, these scientists took on new research problems. Wartime experience had demonstrated the importance of physical oceanography and underwater acoustics for submarine warfare, and in contrast to the more holistic marine science of the interwar period, the oceanography that emerged after the war intensified its focus on physical properties of the ocean.[8]

Postwar American oceanography was characterized both by increased military support and by oceanographers' efforts to once again establish networks of international scientific collaboration.[9] Marine science in the twentieth century became an important currency of diplomacy between the United States and the Soviet Union, even as it gained increased civilian prestige. Capturing the public imagination, oceanography of the postwar period not only enrolled the efforts of private industry but also drew on the expertise and imagination of civilian amateurs. Their participation shaped the course of postwar marine science and ocean engineering.

Scuba Divers: Becoming Part of the Medium

As Marita Sturken and Douglas Thomas have observed, "Technologies take on a special kind of social meaning when they are new." They may "become the focus of intense political, economic, cultural, and even emotional investment.... It is almost inevitably a field onto which a broad array of hopes and fears is projected and envisioned as a potential solution to, or possible problem for, the world at large."[10] This pattern is apparent when we examine the arrival of scuba technology in the 1950s. The growing popularity of scuba diving transformed the oceans in the popular imagination from a dangerous and unknown space into a world of wonder. Self-trained divers saw in scuba opportunities for adventure and for new forms of employment in salvage and underwater construction. John E. Crayford's *Underwater Work: A Manual of Scuba Commercial, Salvage and Construction Operations* (1959) exemplified the trend. It addressed "the young man still deciding upon a career." Crayford provided instructions for everything from underwater demolition to cinematography and biological specimen collection, which, Crayford noted, might find a large market because there were "thousands of high schools, junior colleges, colleges and universities throughout the United States and Canada without access to the sea."[11] The entrepreneurial scuba diver could become an "underwater craftsman," to borrow the title of one of Crayford's chapters.[12] But participation in scientific research offered commercial divers even more attractive opportunities. "Marine research is one of the finest fields which a Scuba diver can enter," Crayford suggested. "Not only is he making a wise decision regarding his career, but he is also furthering man's knowledge of the world in which he lives."[13]

Even before Jacques Cousteau and Émile Gagnan's aqualung was first sold at René's Sporting Goods in Southern California in 1949, free divers had already begun exploring aquatic realms on the western coast of the United States. The world's first diving club, the San Diego Bottom Scratchers, was founded in 1933, and before the invention of wetsuits and swimming fins, "bottom scratchers" used diving goggles and snorkels to spear fish. Spearfishing caught on, and diving clubs and "diving derbies" sprouted along the western seaboard.[14]

As amateur sportsmen turned to the sea, they began to experiment with the improvement of diving equipment.[15] Like radio amateurs, postwar divers benefited from the massive amount of surplus army equipment that had become available for purchase. Two graduate students at Caltech, Howard Teas and Wheeler North, bought oxygen equipment used by air force bomber pilots from an army-navy surplus store, modified it for underwater use, and tried it out in a public swimming pool with mixed success. Teas continued to tinker the following year, using a modified airplane landing gear hydraulic compressor to inject compressed air into diving tanks of his own invention. One of his tanks exploded, nearly taking Teas along with it.[16] The invention of Cousteau and Gagnon's aqualung led to greater standardization of diving equipment, and with the arrival of the wet suit in the early 1950s ever more people turned to underwater exploration. The sport of scuba diving was born.[17]

Some of the new divers were amateur recreationists, but others— primarily, but not exclusively, men—approached diving as a potential paying occupation.[18] Diving in the service of science promised to offer one career path. The Scripps Institution of Oceanography was the first academic institution in the United States to fully take advantage of the new equipment and to add scuba divers to the semipermanent staff. "Do not dismiss the Scuba apparatus as a mere modification of the old diving suit," cautioned Scripps director Roger Revelle. "With it man becomes part of the medium."[19] The first formal training course for diving in the United States, led by diver Conrad "Connie" Limbaugh at Scripps, trained and certified divers for all campuses of the University of California.[20] With the help of the aqualung, graduate students explored California kelp forests and made firsthand observations of marine life.

From Aquatic Ape to Homo aquaticus

In an article in *New Scientist* in March 1960, the British marine biologist Sir Alister Hardy suggested that an ancestor of modern hominids had been forced by terrestrial evolutionary competition to adapt once again to life in the sea. As evidence, he cited various contemporary aquatic mammals thought to have terrestrial ancestors such as whales, seals, and manatees. Some hominids, he claimed, might have been subject to the same evolutionary pressure. And if such an event had happened in the past, it could, and would, happen again. "No one can doubt that history will repeat itself and Man will be forced once again into the sea for a living."[21] Hardy's article was accompanied by drawings of scuba divers corralling fish with a "future submarine tractor trawl." He claimed humanity's future hinged on the successful exploitation of marine resources, primarily marine foods. And he was far from alone in these speculations. In the aftermath of World War II, as the Allied powers struggled to care for thousands of displaced persons in Europe, marine algae attracted attention as a potential miracle food that might solve the problem of global hunger. The Carnegie Institution, presided over by the former head of the United States' wartime science research program, Vannevar Bush, championed the idea.[22] Some people, even more ambitiously, speculated that humans would find ways of diverting ocean currents to increase food production in arid and frozen regions of the globe.[23]

Although evolutionary anthropologists lambasted Hardy's theory of an ancient aquatic ape, scuba divers and marine engineers were prepared to embrace his ideas. Jacques Cousteau deserves much credit for popularizing the idea that the underwater world was destined to become ever more accessible, a message spread through his popular writing and documentary films.[24] Cousteau, in the words of an American journalist, made "his alien environment as familiar to Americans as the *Laugh-In* [a comedy television show] cocktail party."[25] In Europe Cousteau gained increased public authority with his appointment as director of the Oceanographic Museum of Monaco in 1957, a position he held for over thirty years.[26] Speaking before the World Congress on Underwater Activities in London in 1963, Cousteau predicted that there would soon be a "conscious and deliberate evolution of *Homo aquaticus*, spurred by human intelligence rather than the slow blind natural adaptation of species." "After living in compressed air habitats for generations, Water People" would eventually even be "born at the bottom of the sea." Future "alteration of human anatomy" would "give man almost unlimited freedom underwater."[27] American scientists at NASA, he claimed, were already working on an "artificial gill" that could be connected to a diver's bloodstream allowing the filtration of oxygen and carbon dioxide and bypassing respiration—an assertion journalists were unable to confirm with NASA.[28]

According to Cousteau biographer Brad Matsen, when several journalists dismissed his predictions as science fiction, Cousteau countered that they were, on the contrary, quite conservative.[29] His prophesies fell within a larger utopian project of expanding human powers and freedoms, eventuating in the creation of a "new man" through the opening of oceanic space. As he later wrote, "I am not a scientist, I am an explorer.... My goal has always been to free Man from the slavery of the surface, to invent ways and means of permitting him to escape from natural limitations, to breathe in an unbreathable element and resist higher and higher pressures. And not only to resist, but perhaps to adapt himself, to move about, react and live within the sea and take possession of it."[30] This larger utopian project of expanding human powers and freedoms, Cousteau announced, would inaugurate a new era for humanity.

During the late 1960s many believed revolutionary social and political change was imminent, and Cousteau was no exception. Soon there would be "undersea parliaments and new nations," he wrote; "poets, architects, and painters would be needed to give expression" to this "new world."[31] From his own work with experimental underwater habitats, Cousteau concluded that these changes would rapidly follow on from the experience of underwater living. He presented his oil-industry-backed Conshelf program in the Mediterranean in a 1964 Oscar-winning documentary, "The World Without Sun."[32] After visiting divers Robert Falco and Claude Wesly during the seven-day Conshelf I underwater habitat experiment, Cousteau reported, "They were at home there. They had taken on a new mentality. It was no more my business. They were staying underwater but I was returning to the top. This ... makes me believe firmly in the new men of the future.... The birth of a new man is in the line of nature."[33] Many years later, speaking at the first Congress of the Association of Space Explorers in 1985, Cousteau recalled that Falco and Wesly, like astronauts, had passed through a "moral gateway that made them see national and tribal disputes as ridiculous, as something mankind must learn to leave behind."[34]

Technological developments in the years immediately following Cousteau's pronouncements seemed promising for such aspirations.

Walter L. Robb, an engineer at the General Electric Research and Development Center in Schenectady, New York, appeared on the cover of *Science News Letter* in November 1964 along with an "aquatic-hamster." Robb had developed an artificial membrane that filtered oxygen from the surrounding water and permitted the hamster to breathe in an enclosed cage submerged in an aquarium tank.[35] In 1967 a journalist, writing in *Popular Mechanics*, who admitted initial skepticism of Cousteau's predictions, reported that designs for artificial gills had now been patented.[36] Scientists, he claimed, were now "seriously" discussing the possibility of implanted artificial gills, and two divers "from an undersea-engineering firm" had volunteered to undergo surgery.[37] Meanwhile, a research group at Duke University led by Johannes A. Kylstra experimented on mice and dogs to demonstrate the feasibility of breathing oxygenated fluorocarbon liquid.[38] Eventually, Kylstra performed one human liquid-breathing experiment with commercial diver Frank J. Falejczyk, who had already performed simulated saturation diving experiments in hyperbaric chambers at the Duke University Medical Center and the Hyperbaric Chamber Unit at the University at Buffalo.[39] Although the experiment proved dangerous and was not repeated, it foreshadowed later liquid ventilation for premature infants.

Man-in-the-sea saturation diving experiments of the 1960s and 1970s cannot be considered in isolation from contemporaneous experimental work in diving physiology. In a 1963 letter from the American entrepreneur-turned-diving-pioneer Edwin Link to Edward H. Lanphier, a medical researcher at the University at Buffalo, Link explained that a lack of funds prevented him from hiring Falejczyk for purposes of human experimentation, but he offered the hope that "in the near future all of us will be on our way towards better times in this research." He suggested that "in another year" they would be "crying to find men like him." Link also experimented with animal subjects. In the same letter he describes simulated dives with mice to the equivalent of 2000 feet and remarks that he had been "tremendously interested" to learn of Lanphier's "oxygenation work using a dog." Finally, he informed Lanphier that his team would soon build a larger pressure chamber "to contain animals as large as goats" at a simulated pressure of 3000 feet, and he requested Lanphier's advice on the most suitable animals for such experiments ("dogs, goats, monkeys, etc.").[40]

By 1964 Cousteau had moved forward with several ocean installations. His subsea Conshelf habitat program was now in its second stage, and emboldened by the commercial success of his documentary films, he began preparing an ambitious third habitat, Conshelf III.[41] That same year he launched a floating laboratory platform in the Mediterranean named the Mysterious Island after the Jules Verne novel of the same title.[42] The floating laboratory provided a platform for in situ marine research, and the press was quick to label the installation a "floating island." Rotating teams of four scientists manned the station, which was positioned approximately halfway between Monaco and Corsica. The carefully chosen site, described by one researcher as "the eye of a liquid cyclone," the calm center of a turbulent gyre, was approximately a hundred miles offshore over water one and a half miles deep.[43] Unlike the free-floating FLIP platform built by Scripps Institution of Oceanography for hydroacoustic research that had been launched two years earlier, the Mysterious Island was tethered to the seafloor but capable of moving freely within a confined radius.

Women and televisions were forbidden on the platform, but with the collusion of some Russian oceanographers visiting from the research vessel *Mikhail Lomonosov*, a television set was smuggled aboard in July 1969. And there, floating in the Mediterranean, a small group of French and Soviet scientists watched grainy black-and-white images of American astronauts first setting foot on the moon.[44] They may have felt some fraternity with the lunar explorers, for both Cousteau's Mysterious Island and Scripps's FLIP—first steps in the colonization of the oceans—promised the dawn of a new age.

Ocean Colonization and Human Survival

Fueled by institutional growth and generous postwar funding, oceanographic science—particularly physical oceanography—grew rapidly in the 1950s, 1960s, and 1970s. At the same time, oceanographers anticipated dramatic advances in ocean engineering. Newspaper editorials proclaimed that humans would be mining, farming, and living on the seabed in the near future.[45] The sea seemed to offer salvation from multiple widely perceived threats to humanity's future. This increasingly articulated confidence in marine salvation was reinforced by the success of widely publicized man-in-the-sea projects, from Cousteau's Conshelf programs in the Mediterranean and Red Sea to the US Navy's Sealab and Tektite underwater habitat projects in California and the Caribbean.[46] Americans, enraptured by the adventure of underwater exploration, were eager to understand the technological feats required for further conquests of the deep.[47]

In the late 1960s, seamounts seemed to offer uniquely promising sites for underwater habitat installation. An article in *Popular Science* claimed that "within 10 years ... it should be possible to assemble a scientific colony thousands of feet below the surface atop the seamounts of the Mid-Atlantic Ridge." With a depth range of up to 15,000 feet, seamounts "could be used as platforms for several kinds of undersea bases." The author of the piece suggested that, as on land, control of the high ground in the oceans was of the greatest strategic military value, quoting a "weapons-oriented scientist" as stating, "If we control the ridges, we can control the oceans."[48]

In contrast to such military preoccupations, visitors to the 1964 World's Fair in New York were offered a peaceful vision of future colonization of the ocean by General Motors' Futurama. A train carried visitors past animated dioramas, including one highlighting the "never-ending nourishment" to be made available through new forms of oceanic farming and harvesting. Abundant food was promised for seven times the human population. In other scenes, "aquacopters" searched the ocean floor for "vast fields of precious minerals and ores," and in shallower waters, visitors glimpsed "Hotel Atlantis," a place "of radiant wonders" and submerged "sun-bright gardens."[49]

Dreams of ocean settlement were not restricted to the popular imagination.[50] Scientists regarded seamounts as intriguing new areas for study, as sites of fascinating flora and fauna, and as practical sites for the deployment of stationary instrumentation in the open ocean.[51] The military carried out studies of underwater construction materials and assessed the feasibility of building undersea installations throughout the 1960s. For instance, a 1966 report by Carl F. Austin, a research geologist at the US Naval Ordnance Test Station at China Lake, California, investigated the possibility of a "rock site" installation, suggesting that living and working spaces could be built in the bedrock beneath the seafloor.[52] Private companies hoping for lucrative military contracts conducted similar studies. General Electric and Corning Glass Works announced plans to build an underwater habitat, Project Bottom-Fix, at a depth of 12,000 feet on the Mid-Atlantic Ridge.[53] The publicity given these plans also raised concerns about the absence of legislation to control and manage development in marine spaces.

In 1967 President Johnson appointed the National Commission on Marine Science, Engineering, and Resources. Members included policy makers, academics, and business executives. In its final report, the commission expressed concern that the use of coastal areas had already "outrun the capabilities of local government" to manage "orderly development and resolve conflicts." The commissioners feared that "procedures for formulating sound decisions" were inadequate.[54] They noted that in the future "recreation diving from underwater habitats and touring in glass bubbles and small submarines" might become the norm.[55] To both encourage and manage offshore development, the commissioners suggested that underwater "seastead" leases, modeled on the 1862 Homestead Act and "contingent upon the useful development of the marine tract in a manner that would safeguard necessary navigation, fishing, and other uses," should be used to capture and properly channel "public interest."[56]

By far the most ambitious national proposal was the commission's recommendation that an "Experimental Continental Shelf Submerged Nuclear Plant," capable of powering "continental shelf laboratories," be jointly funded by the National Oceanic and Atmospheric Administration (NOAA) and the Atomic Energy Commission. Continental shelf laboratories were necessary, the commissioners claimed, because "if man is to conquer the sea, he must go into the sea."[57] Their message was clear: to fully access marine resources the oceans had first to be colonized in a well-planned manner.

Competing Claims over Marine Space

In the late 1960s, just as Project Sea Use was getting underway in Washington State, an international legal debate raged over who could lay claim to ocean space. Maritime law had remained largely static since the late seventeenth century. It was then that the Dutch jurist Hugo Grotius published his famous *Mare Liberum* from which was derived the concept of freedom of the seas: the open sea, initially defined as beyond the distance of a cannon ball fired from the coast, lay outside the territorial jurisdiction of any nation. Not until the nineteenth century did the exploitation of maritime resources motivate a reassessment of maritime territorial law. In northern Europe, coastal states attempted to prevent overexploitation of fisheries by establishing quota limits. These attempts failed because there were no ways to enforce compliance. In the aftermath of World War I, the League of Nations attempted, but failed, to secure international support for a three-nautical-mile territorial limit.[58] At stake for the United States was the Coast Guard's ability to detain smuggling vessels along the Atlantic Seaboard beyond the three-mile limit.

A decisive event occurred on September 28, 1945, when President Harry S. Truman issued Proclamation 2667, declaring US jurisdiction over the "natural resources of the subsoil and sea bed of the continental shelf." While laying claim to geologic resources in the continental shelf, notably offshore oil deposits, Truman's proclamation also preserved "free and unimpeded navigation" of the "high seas of the waters above the continental shelf." Partly in response to this extension of American territorial claims, Chile and Peru declared a two-hundred-nautical-mile territorial exclusive economic zone to protect their fisheries. Costa Rica, El Salvador, Honduras, and Ecuador followed suit, despite US protests. Tensions rose in subsequent years, leading to confrontations between the Peruvian navy and American commercial fishing vessels.[59] In 1953, the United States passed the Outer Continental Shelf Lands Act, extending territorial control over seabed out to three nautical miles (5.6 km) from the coast. This legislation extended the "Constitution and laws and civil and political jurisdiction of the United States . . . to all artificial islands and fixed structures which may be erected thereon for the purpose of exploring for, developing, removing and transporting resources therefrom, to the same extent as if the outer Continental Shelf were an area of exclusive Federal jurisdiction located with a State."[60]

These events, and subsequent declarations of two-hundred-nautical-mile-limit claims by other nations, led to the first United Nations Conference on the Law of the Sea in 1958. But neither this nor a subsequent meeting in 1960 resolved the issues at stake. "Legal chaos now reigns at sea," wrote a *New York Times* journalist in 1963. "Most nations are now making their own rules for their surrounding waters, pursuing rival economic and defense objectives with arbitrary claims of sovereignty."[61]

Whereas some looked with anxiety to legal uncertainty over offshore sovereignty, others recognized opportunity. In Europe, so-called pirate radio stations began popping up on ships and platforms in the late 1950s and early 1960s. By operating just beyond national territorial jurisdictions, they were able to evade broadcasting regulations.[62] In the United States, some entrepreneurs dreamed bigger. In 1965, a "Louisiana investor" named Louis M. Ray imagined founding a new country offshore. He would build it from reclaimed sections of the Triumph and Long reef only a few miles southeast of Miami and call it the "Grand Capri Republic." Although Ray succeeded in attracting investors for a proposed $250,000,000 casino and preliminary construction work began, the venture was halted by a Miami-based federal judge. In his final ruling, Judge Charles Fulton argued that the reef could not be claimed because it was part of the seabed and constituted a natural resource, as defined by the Outer Continental Shelf Lands Act of 1953. Fulton's ruling cited possible environmental destruction of the reef but also opposed the legal precedent the venture would establish. At risk, Fulton claimed, was "the preservation of our very security as a nation." Ray lost his appeal, the final ruling declaring that his "dreams must perish, like the lost continent Atlantis, beneath the waves and waters of the sea."[63] If offshore areas were colonized and placed outside the control of the US government, they could conceivably support "not only artificial islands and unpoliced gambling but even alien missile bases."[64] The legal management of marine territories had become a matter of national security.

In 1968, the National Petroleum Council, a powerful political lobby, entered the arena after a series of reports by the US Geological Survey claimed that seabed oil deposits were predominantly on the continental margin. Estimates for the US continental margin predicted yields of up to 780 billion barrels of oil and 2,200 trillion cubic feet of natural gas.[65] The council urged the federal government to take control of the continental margin, both to safeguard national security interests by minimizing dependence on foreign oil and to stave off a looming national energy crisis. This proposal ran counter to the interests of the Department of Defense, whose primary concern was that foreign powers might take advantage of this precedent to install antisubmarine detection instruments on the seabed off their shores.[66] The Defense Department feared that limited foreign national-resource sovereignty might eventually become total territorial sovereignty through "creeping jurisdiction" and, by extension, would eventually limit freedom of movement at sea.[67] While a change in policy might benefit US corporate interests, Defense argued, 92 percent of the world's continental margins were off foreign shores, and it was this larger consideration that mattered for American national security interests. Instead, Defense proposed the establishment of an impartial international seabed authority, an organization with which the petroleum industry could more easily negotiate than with unpredictable national governments.[68] The American petroleum industry, primarily concerned with limiting foreign access to oil fields off US shores, successfully lobbied against the Defense Department, persuading the federal government to lay claim to the continental shelf.[69]

The US mineral extraction industry too had vested interests in the marine jurisdiction debate. Initially siding with the petroleum industry, mining lobbyists came to favor a regime of international governance once it became apparent that potentially profitable manganese deposits lay beyond the continental margin. Shifting position, they lost allies in the petroleum industry but then also found themselves in opposition to the Defense Department, which was advocating an international seabed authority supported by offshore mining revenue. With support from neither the petroleum industry nor the Defense Department, the mining industry was marginalized in the policy debate.[70] Soon, actions of other nations forced the State Department's hand. Brazil laid claim to a two-hundred-nautical-mile territorial sea in 1968 and Canada declared a one-hundred-nautical-mile jurisdiction extension under its Arctic Waters Pollution Prevention Act in 1970.[71] The US government eventually renounced seabed claims beyond a depth of two hundred meters and called for the creation of an international regime to govern seabed resources beyond that point.[72] It advocated a "trusteeship zone" managed by coastal nations on behalf of the international community extending from two hundred meters' depth to the edge of the continental margin. Each coastal state was to receive a share of the revenue generated within these zones.[73]

Marine scientists, through their professional bodies, strove to ensure continued access to the ocean basin in its entirety. The 1957–1958 International Geophysical Year (IGY) had demonstrated the value of international collaboration for tackling oceanographic problems. As part of the IGY effort, the International Council for the Exploration of the Sea organized an extensive study of the North Atlantic using twenty-two research vessels and enlisting eight nations (see Figure 4.1). The result was the most complete survey yet made of the boundary zone between the Gulf Stream and the Labrador Current—crucial for meteorological and fisheries research. The success of the IGY effort led to the 1959–1963 International Indian Ocean Expedition, involving forty research vessels, and to the establishment in 1961 of the Intergovernmental Oceanographic Commission, a sub-branch of UNESCO.[74]

Legal territorial jurisdictions posed an impediment for such cooperative expeditions. In June 1968 the Scientific Community on Oceanic Research of the International Council of Scientific Unions issued a statement decrying the impingement of maritime law on oceanographic research: "Evidence is accumulating that the Convention on the Continental Shelf, not ratified by many maritime nations, is on occasion being applied so as to hinder scientific investigation of the circulation of ocean waters, the biology of the sea floor, the origin and movements of continents, and other problems of considerable scientific importance." It urged its membership to press their governments to "adopt liberal interpretations of the articles of the Convention in order to facilitate the carrying out of oceanographic research."[75] The first proposal for an international agreement for safeguarding the communal interests of oceanographers originated in the Soviet Union and was presented at the 1967 meeting of the Bureau and Consultative Council of the Intergovernmental Oceanographic Commission. The call was taken up by the National Commission on Marine

Figure 4.1 One of the posters for the 1958 International Geophysical Year. National Academy of Sciences, *"Planet Earth": The Mystery with 100,000 Clues* (Washington: National Academy of Sciences), 10.

Science, Engineering, and Resources in a 1969 report, "Our Nation and the Sea." It was a plea for freedom of scientific research "in the territorial waters or on and concerning the continental shelf of a coastal nation" so long as the nation was notified.[76]

While these issues were being raised, some analysts argued that the debate failed to account for seamounts: terrains classifiable neither as continental shelf nor as deep sea. Yet it was precisely this terrain that was of strategic military and commercial importance. In 1968, the editors of *Oceanology International* asked, "Who owns the Cobb Seamount?" The legal answer, they concluded, was "nobody." Yet they had to concede its "potential for scientific research [was] obvious" as was its "potential for military exploitation." According to the same article, the Naval Undersea Warfare Center had already completed a "preliminary evaluation of the seamount for possible use as a submerged military base." With Project Sea Use going ahead, they noted, "once Americans have occupied the Cobb Seamount, even for such a short period of time, the next logical question is whether this constitutes a valid claim to the area." Existing diplomatic treaties suggested it would not, unless "loopholes" could be found.[77]

A 1970 paper published in the *Marine Technology Society Journal* sought to highlight the legal issues posed by seamount use: could the United States legally build a scientific outpost in the "High Seas"? The answer, asserted its author, was "germane not only to the Sea Use Program" but also to similar operations "now being conducted or which will be conducted in the future by the United States or its Citizens."[78] Also thrown into question were jurisdictional boundaries between state and federal powers. Writing in 1970, Frances and Walter Scott noted, "With the states claiming the seabed under territorial waters and the federal government owning the continental shelf, it is uncertain what governmental agency would have the power to encourage or restrain the organizations about to occupy a seamount. As new technological developments such as submersibles and habitats make exploration of seamounts possible, the Law of the Sea will be affected by every event on the ocean floor."[79]

Man-in-the-sea efforts of the late 1960s and 1970s should also be understood in the context of Cold War competition over marine territory. In a 1971 study of Soviet adherence to international law of the sea, the legal analyst William E. Butler concluded that although both the Soviet Union and the United States had vested interest in the militarization of the seabed, there was reason to doubt that Soviet capability matched that of the Americans.[80] Soviet saturation diving research during the Cold War remains understudied, but we do know that at least ten different Soviet habitats were constructed between 1965 and 1972. Others were built in Soviet satellite countries: Bulgaria, Czechoslovakia, Poland, Romania, and Cuba.[81] US intelligence assessments at the time suggested that Soviet saturation diving research lagged behind. A 1973 NOAA report concluded, "The Soviet diving programs are behind those of the United States and European nations. . . . Soviets are interested in western MUA [manned undersea activity] technology and have attempted to purchase US and Canadian DSV's [deep submergence vehicles] over the past few years." From what is now known it appears that Soviet-built habitats were limited to shallow water, not having been designed to withstand depths exceeding two hundred meters.[82]

It seemed clear by the early 1970s that Soviet saturation diving did not pose a military threat, but during the 1960s the US Navy was unwilling to risk a strategic disadvantage.[83] At a time when army and air force engineers were drafting proposals for a lunar military base, the navy began feasibility studies for seafloor bases.[84] Carl F. Austin of the US Naval Ordnance Test Station at China Lake in California, one of the first researchers to work on the problem, noted, "Using only the tools and techniques [of] today's raw materials industry, manned installations of a large size containing a one-atmosphere shirt sleeve environment . . . can be established at almost any location on the continental slopes, the deep-ocean floor, and on seamounts and ridges."[85] In a 1967 report of the ocean engineering program, the navy argued that sub-bottom structures offered potential as "laboratories, military support facilities, recreational facilities, and industrial installations."[86] One year later the *Interim Report of the Navy Deep Submergence / Ocean Engineering Program Planning Group* included among its list of objectives "construct installations on the seafloor," "support bottom installations," "police / protect bottom installations," and "neutralize enemy undersea installations."[87]

Nevertheless, the ambitious scope of such plans proved short-lived. The navy's 1968 Project Blue Water report stated that "all efforts underway in the Man-in-the-Sea project concern the continental shelf aspect; the deep ocean aspect, defined in a separate Advanced Development Objective, is not being pursued at the present time."[88] The reason for the truncated scope, as one oceanographer suggested to a reporter in the summer of 1969, was that there had never been an "oceanographic equivalent of Sputnik."[89] And without a boost of federal funding, NOAA could never attain the status of a "wet NASA."

The Mountain under the Sea

Having sponsored the ambitious Marine Sciences and Research Act of 1961 Senator Warren G. Magnuson of Washington State gained the nickname "Mr. Oceanography."[90] In a section of the act titled "Marine Science—the Neglected Frontier," he wrote that the ocean was "the earth's largest laboratory" and the Great Lakes the "earth's largest fresh water laboratory." To study these areas would require "a national effort and program," because it was a task that no single state or government agency could accomplish. But in pleading his cause he did not argue that a national program for the study of the oceans was imperative for the sake of science alone. Rather, the threat of Soviet supremacy on the seas was cited as a spur. "Soviet research vessels carry a bronze plaque of Nicolai Lenin with this quotation: 'In order to spread world communism, it is necessary to use the fields of science and technology.'" The seas, he warned, are crucial in the Kremlin's "scheme of world domination."[91]

After an incident in which a Soviet trawler rammed an American fishing vessel off the Washington coast, Magnuson, then chairman of the Senate Interstate Commerce Committee, demanded reparations from the Soviet government, describing the event as "bordering on an attack." He warned that similar incidents of such an "explosive nature" would likely be repeated if the Soviet fleet remained in the area. "The U.S. position toward the Soviet fishing fleets," he advised the US fishery delegation then meeting with the Russians in Moscow, must be "a forceful one."[92]

Cobb Seamount is a relatively unusual marine feature because its summit, only twenty-six feet from the surface, lies within the photic zone—the depth to which sunlight can penetrate. The surface of the summit has an area roughly equivalent to that of seventy-eight football fields. Farther downslope, the base of the seamount spreads out into a terrace at what has been described as "twilight" depth. Cobb is 270 nautical miles west of Gray's Harbor, Washington—far enough from the shore to be in international waters. Cobb has the shallowest summit of any seamount off the western seaboard of the continental United States. Because of its proximity to the surface and the localized upwelling generated by the sloping topography, the area is a rich fishing ground.

Cobb Seamount was first discovered in the summer of 1950 by Bureau of Commercial Fisheries vessel *John N. Cobb*. Its abundant fauna and unusual geology attracted the attention of regional research institutions, notably the University of Washington, the University of Oregon, the University of Victoria, the Earth and Space Sciences Agency (predecessor to NOAA), and the Bureau of Commercial Fisheries (predecessor to the National Marine Fisheries Service). Investigation of Cobb as a possible open-ocean instrument platform began in earnest only in 1960 when Robert G. Paquelle of the University of Washington Oceanography Department obtained $207,000 in project funding from the National Science Foundation to install a "telemeter buoy" on its summit.[93] Newspapers reported that the mountain would serve "as an anchor for a scientific instrument designed to probe the secrets of the deep."[94] Shortly thereafter Cobb came to the attention of Honeywell, then operating a development laboratory in Seattle. In a letter to University of Washington's director of the Department of Oceanography, Richard Fleming, Theodor Hueter, manager of the Honeywell Seattle laboratory, expressed company interest in "instrumentation for high-pressure environments, underwater acoustics and telemetering and data handling," proclaiming the company's desire to "contribute toward a successful attack on the Sea Mount task."[95]

With this university-industry partnership secured, on June 19, 1965, the Associated Press reported that five divers from the University of Washington "had reached the summit of a 10,000-foot peak under the surface of the Pacific." The summit was being investigated as a possible base for a radio tower to transmit "oceanic scientific data." The tower was meant to prove that an "unmanned data collection and transmission system" could be successfully installed in the open ocean. It would be, claimed Walter Sands, Fleming's successor as director of the UW Department of Oceanography, "the only one of its kind on the high seas."[96]

The Oceanographic Commission of Washington (OCW), established in 1967 by state legislation as a "policy group, with advisory, promotional and educational duties," comprised twelve commissioners drawn from state legislature and administration, industry, labor, and education, and it employed a permanent staff of four administrators.[97] A second organization, the Oceanographic Institute of Washington, was created at the same time. The institute, a nonprofit organization composed of all members of the commission plus an additional eight elected members, served as the "action arm" of the OCW. In an opinion piece Senator Gordon Sandison of Washington argued that the institute and OCW would "encourage, assist, develop and maintain a coordinated program in oceanography and oceanographic research for the state's citizens and for the nation."[98]

With this body in place, in 1968 several research institutions—the University of Washington, Battelle Northwest, and the Oceanic Foundation of Hawaii—in cooperation with the private marine engineering company Honeywell Marine Systems, petitioned the OCW to sponsor a coordinated research program for Cobb Seamount.[99] With the backing of the commission, Project Sea Use was born.[100] Speaking before Congress in May of that year, Senator Magnuson described the location and features of Cobb Seamount, arguing that the project would "establish a series of important precedents which should further the ocean sciences throughout the country." Such precedents included "the first integrated employment of modern deep ocean technology in direct support of subsea scientific research; the first manned habitation of an ocean seamount; the first definitive scientific exploration of a subsea site by man living within the environments." Emphasizing the strategic importance of this effort, he observed that "ownership and control of the ocean and its resources beyond the continental shelves" remained an "unanswered question." It was, therefore, in the interest of the United States to "occupy Cobb Seamount." "Should it be occupied by another nation it could be an important strategic loss for our country." On an idealistic note he concluded that Project Sea Use would "serve as the major U.S. initiative in the development of inner space for the benefit of mankind."[101] Senator Henry M. Jackson, also from Washington, of the Senate Armed Services Committee lent his support, contending that Cobb could eventually serve as a "base for long-range ocean surveillance systems."[102]

Magnuson and Jackson were not alone in pointing to the strategic value of seabed occupation. In 1966 the chief scientist of the Special Projects Office of the Navy, John P. Craven, writing in the *U.S. Naval Institute Proceedings,* suggested that "for the greater portion of the continental shelf (0 to 3000 feet) the economic use of man and machines for extensive and prolonged engineering operations is virtually assured within the next decade," as was "the capability to deploy vehicles and to mount installations" on ocean ridges and seamounts. Foreseeing that the law of the sea had rapidly been adapted to grant territorial rights over areas of the continental shelf, Craven confidently stated that "the ability to exert sovereign rights in the entire sea bed has already received tacit approval."[103]

Provision of underwater infrastructure was never a top priority for the Johnson administration, and the president's naval advisors were forcefully reminded of the need to give attention to underwater capabilities when the US nuclear submarine *Scorpion* sank in the North Atlantic on May 22, 1968, with ninety-nine lives lost. The sinking of the *Scorpion,* following an earlier sinking of the nuclear submarine *Thresher* in April 1963, confirmed naval strategists' fears that American marine technology remained ill-equipped to deal with problems of deep ocean survey, salvage, and rescue.[104]

Meanwhile in Seattle, Project Sea Use was the key item on the agenda of the fifth meeting of the OCW in April 1968. The project was conceived as "a long-range program to beneficially utilize Cobb Seamount," and the minutes of that meeting list its objectives:

1. Demonstrate that a region can initiate, organize and execute a proactive scientific program involving an important regional resource with broad participation from educational institutions, private industry and the State and Federal governments.
2. Characterize the chemical, physical, geological and biological features of the Seamount and its environs and plan its future utilization for continued benefit to the Northwest.
3. Demonstrate that man can occupy a seamount, perform meaningful scientific work and perform underwater construction at a seamount, far distant from land based support and facilities.

4. Use presently available deep ocean technology in integrated support of a scientific program.
5. The ultimate and potentially most important result is concurrence that *Project Sea Use* is a desirable, cooperative accomplishment, designed to result in a detailed analysis and recommendation concerning continuing and ultimate utilization of the Cobb Seamount site.[105]

In May 1968, the press announced the launch of two underwater habitat projects. The first, Tektite, would house four scientists for sixty days—the longest continuous undersea stay yet attempted—in Greater Lameshur Bay in the Virgin Islands (see Figure 4.2). The Tektite project was a collaborative venture of the navy, NASA, and the Department of

Figure 4.2 Artist cutaway painting of the Tektite II habitat. NOAA Central Library Historical Fisheries Collection / fish / 9765.

the Interior, and the habitat construction was contracted to General Electric's missile and space division. The name Tektite reflected a dual mission, to study the marine environment and to simulate conditions of long-duration space flight.[106] A second project announced in the same month was a "$2.1 million expedition . . . that would send aquanauts to live in an undersea laboratory on a volcanic plateau in the Pacific Ocean." Cobb Seamount was billed as "an ideal place to study ocean floor geology at first hand."[107] Tektite proceeded on schedule, but budget constraints hampered the Cobb program.

Project Sea Use involved the collaboration of four major actors: state and federal legislators, academic scientists, leaders of private industry, and a group best described as diving tinkerers. I use the word "tinkerers" not to connote a naïve or amateur approach to the work but rather to highlight that these civilian divers were, to a large degree, self-taught adventurers who relied on improvised instruments and diving procedures using local resources and materials. The project was made possible only by the labor and technical expertise of this group.[108]

The Washington legislature was reluctant to give up on the dream of a habitat program in the Pacific Northwest. Appearing before the House of Representatives Subcommittee on Oceanography, Conrad Mahnken, veteran aquanaut of the Tektite program, recommended Puget Sound as a site for "a future *Tektite*." Mahnken testified to the interest in a habitat such as Tektite of scientists in the Puget Sound region because they hoped "to place an underwater habitat on Cobb Seamount . . . [to] be used primarily as an observational platform for measurements in the open ocean." Mahnken cautioned that the technical and scientific expertise required for the venture would have to be developed in a "shallower and more protected environment." In Puget Sound, such a habitat could be located at a depth of up to one hundred feet (thirty meters) and serve as "an underwater platform for technicians and scientists for studying underwater pens of salmon or underwater rafts of mussels and oysters." Asked whether he was speaking on only his own behalf, Mahnken answered that his recommendation was supported by fellow aquanaut John VanDerwalker and Washington congressman Thomas Pelly. Mahnken and VanDerwalker were employees of the Bureau of Commercial Fisheries.[109]

Following the successful completion of Tektite I, Congressman Pelly and Congressman Robert Sikes from Florida, petitioned to have the Tektite project relocated from the Virgin Islands to their states. Some involved with the project resisted. As one *"Tektite* man" informed the *Virgin Islands Daily News,* "A lot of people who aren't knowledgeable think you can pick up the Habitat and put it in any water." A move to cold water would require extensive modification or a complete redesign.[110] A 1969 naval report concluded that marine terrain near Hawaii offered the most promising location for future ocean construction experiments.[111] Ultimately, Tektite remained in the Virgin Islands, and the scientific work it supported remained focused on tropical reef ecology.

Despite these setbacks, a habitat was still scheduled for deployment on Cobb in 1969. A habitat already under construction by the Pittsburgh–Des Moines Steel Company for the Oceanic Foundation in Hawaii was marked for relocation to Washington.[112] At 300,000 pounds, it consisted of three isolatable and connected pressure vessels mounted on catamaran hulls and rated for six hundred feet (183 meters) of depth.[113] But back in Seattle, by July of 1968, members of the OCW were noting the failure to generate financial support to run a habitat program. As reported in meeting minutes, "Solicitation for federal support has not turned up any likely funds. As a result it has been necessary to put together a volunteer effort in order to keep this project moving." They were unwilling to completely abandon the plan and hoped that the habitat might still be set up by 1970 or 1971.[114]

The most complete account of the proposed scope of Project Sea Use is a report produced by the Battelle Memorial Institute. According to this 1968 document, Cobb Seamount, surveyed by the Naval Undersea Warfare Center, was judged ideal "for a possible 'Manned-in-Bottom-Base' or MIBB." The report added that "from a technological standpoint, there appears little doubt that such a concept as MIBB can be achieved."[115] The project secured at least unofficial endorsement from the navy. At a meeting of the Undersea Technology Industry Clinic, Vice Admiral Turner F. Caldwell Jr., director of US Navy Anti-Submarine Warfare programs, stated that Cobb Seamount "would furnish an excellent means for developing legal concepts of utilization and occupation of real estate at the sea floor."[116] The secretary and default leader of the Sea Use program,

Emory Day Stanley Jr. told journalists that Cobb was "a resource not duplicated or available in any other region of the United States," expressing his hope that it might become the first "national seamount station." Funding was still lacking, but Stanley was undaunted: "When we do go to Washington, they'll hear us coming and, we hope, be convinced of the validity of what we're going to do."[117]

Sharon Dodge, a Washington native who had gained regional media attention as a backup diver in the all-female aquanaut team at project Tektite in the Virgin Islands, told a Seattle reporter that the Pacific Northwest was the next frontier for underwater exploration: "So many people noted in the field are moving here—you can almost smell something big in the air." She informed the paper that she and her husband had already applied to participate in a two-person "minitat" habitat program then being tested in the Virgin Islands that, she hoped, might be relocated to Puget Sound.[118]

The international legal ramifications of setting an "occupation" precedent on Cobb did not escape public notice. As a staff writer for *Science* noted in a July 1968 editorial, "Although there are defense-oriented groups within government that wish to see Cobb Seamount and other such sea formations close to American soil under U.S. jurisdiction, official government policy has been one of 'open occupancy.'"[119] Such concerns, however, had few advocates.

The Sea Use Aquanauts

Although the navy was never directly involved with the Sea Use program, it did provide technical support and funding through the Office of Naval Research and the Naval Civil Engineering Laboratory. As with the Scripps Institution of Oceanography, the faculty of the University of Washington Oceanographic Department enjoyed a close relationship with the navy. For example, in a 1963 letter to the head of the University of Washington Oceanography Department, Richard Fleming, Rear Admiral E. C. Stephan, chairman of the navy's Deep Submergence Systems Review Group, requested a review of the department's "present research capabilities as they might relate to problems of location, detailed observation, rescue and salvage of submarine-like objects" and offered hope that naval research might produce salvage technology useful "in other phases of marine science and engineering." In his reply to the admiral, Fleming reported that his department had an ongoing investigation of "the feasibility of an open-sea submerged buoy, with a three-point moor and surface-piercing mast, for use on the continental shelf." Regarding "ocean engineering," he noted that it was still "ill-defined" and "as yet only borrow[ed] techniques from other fields." This might change, he suggested, "as increased national effort builds up a broader base of competence and interchange of experience."[120]

The navy had an informal tie to Project Sea Use through Sea Use Council secretary Stanley, a retired rear admiral, and its project director, Charles L. Gott, a retired navy commander. The civilian institutional collaborators were Honeywell's "deep sea research laboratories," the private nonprofit Battelle Memorial Institute, the University of Washington, and the Oceanic Foundation based in Honolulu. Washington's governor, Daniel Evans, wrote to the governors of Hawaii, Alaska, and Oregon inviting wider regional participation in the project. However, according to the July 1968 meeting minutes of the OCW, only Governor John Burns of Hawaii responded. Despite the OCW's hopes that Alaska and Oregon might become involved "at the working level," they appear to have declined.[121]

The man given the task of assembling and leading the aquanaut team was Spence Campbell. After leaving the air force, Campbell was looking for work and knew that he loved being in the water as much as he loved flying. At age twenty-six he swam from Idaho to Oregon, down the length of the Clearwater, Snake, and Columbia rivers, retracing the route taken by Lewis and Clark (in boats) and beating their time by three days.[122] Inspired by books on underwater photography and by the fictional television adventures of actor Lloyd Bridges in *Sea Hunt,* Campbell sought adventure. In 1959 he enrolled in the Coastal School of Deep Sea Diving in Oakland, California, an institution that ran ads for prospective students in diving magazines, calling deep-sea diving "next to space travel, the most exciting career in this man's world" and urging readers to "be an underwater pioneer." Shortly thereafter Campbell moved to Spokane, Washington, where he worked as a commercial diver on hydroelectric dam construction.

When Project Sea Use began, Campbell was working at the Virginia Mason Medical Center in Seattle as a collaborator of vascular specialist Dr. Merrill Spencer, founder of a research lab to study diving physiology. Spencer hired Campbell, who had been conducting his own decompression experiments in a hyperbaric chamber assembled in his garage, to manage the lab. The new team began a series of experiments to test the navy's "exceptional exposure" diving tables.[123] This involved hyperbaric chamber simulation dives and decompression ascents using sheep, goats, pigs, and rabbits.[124] These experiments revealed the navy's diving tables to be faulty. Navy tables suggested a first decompression stop at 18 meters from the surface after a 60-meter dive, whereas the lab discovered that vascular gas emboli began to appear at 30 meters down; a diver ascending to 18 meters would sustain damage to the vascular system that was no longer treatable. To monitor gas buildup in the bloodstream, the team developed a new detection technique using ultrasound, first by implanting probes in experimental animals and eventually designing a unit capable of monitoring through the skin. Blood bubbles passing through vessels produced a distinctive clicking noise picked up by the ultrasound monitor. This new technique became essential for the diving work eventually done on Cobb, in which precordial (chest region) monitors detected gas bubbles building up in a diver's bloodstream. Using these monitors, Sea Use divers were eventually able to entirely eliminate incidents of the bends.

When Sea Use divers entered the water, they discovered an otherworldly landscape. One diver was reminded of the surface of the moon. "My brain was so full with what I was seeing," recalled another, "it was hard to make my mind work."[125] The steep-sided seamount extends upward from murky depths. Marine plants growing in cracks outline polygonal patterns on the rocky surface, and dense schools of fish (giant cod, red snapper, groupers, and the odd sun fish) swirl through, illuminated by sunlight reflecting from the scoured white rock below. Larger predators lurk there as well: lion's mane jellyfish with tentacles five meters long and sharks. As Campbell told me in an interview, "Every vertical crevice down there is loaded with tube worms and rock scallops."[126] In one instance a thirty-pound rockfish stole a crescent wrench. On another occasion, as Campbell and another diver stopped on their ascent for decompression, a large great white shark appeared, passed close by, and then charged straight toward them. The divers made an emergency ascent, quickly grabbed two spare scuba tanks from their Zodiac, and dove back into the water to decompress, opening the tank valves and creating a noisy curtain of bubbles until the shark turned away.

The team of divers for the Cobb mission had to be carefully chosen, with fearlessness one of the primary selection criteria. As Campbell told me, "There's no way to describe, even though a guy is a really excellent diver, and has a ton of experience, and is a tough young guy, the feeling of jumping off of that ship, and into that ocean when you can look around and see nothing but 360 degrees of ocean around you and then you wonder what's down there. I had a diver who literally got down on his hands and knees and cried and couldn't get off the ship. And he was a good diver! It scared him that bad."[127] Working in the Virginia Mason Medical Center lab was also nineteen-year-old Roland White, who soon joined the diving project. "I was just a crazy teenager, so I would do anything," he told me.[128] The aquanauts on the first mission represented three organizations, the Virginia Mason Medical Center, the University of Washington, and the Honeywell engineering test department.[129] In a scrapbook Campbell kept to document the Sea Use program he wrote, "I picked and specially trained the divers for this mission. I readily admit that I was not as experienced as others for this size and scope of expedition; the opportunity was given to me and I wanted the challenge! . . . The seamount is an unforgiving teacher!!"[130]

Serving as advisor to the diving team was Jon Lindbergh, a veteran aquanaut of Edwin Link's Man-in-Sea program and son of famed aviator Charles Lindbergh. The chief scientist for Sea Use I was Robert Burns, director of the Environmental Science Service Administration's joint oceanographic research group at the University of Washington. Before the divers headed out to the seamount, simulated dives were carried out in a diving chamber to weed out anyone susceptible to oxygen poisoning. On the seamount the divers would breathe a gas mixture of nitrogen and oxygen, allowing them to work for a longer time at depth. Despite these precautions, diving conditions warranted apprehension.

During the first mission, divers Chuck Blackstock and Jim Gavin did a line-tended dive with a hose. The team was under the false impression that they could tend a line within five degrees of vertical. Gavin's air supply hose was accidentally cut off, and he suffered an air embolism and was unconscious when pulled aboard. After being moved to a shipboard decompression chamber, he recovered. But the mission was cut short. The dive team also required regular treatments for joint pain caused by gas bubbles forming in their blood.

Initially, it was difficult for the vessel to stay on station above the seamount. After a failed attempt to use explosive anchor bolts designed by the naval office of civil engineering, this problem was overcome using anchors cemented into the rock. Divers would swim down to attach the ship's mooring line to the anchor bolts, and the ship could then pull in the slack. Campbell carefully documented all missions in a scrapbook of news clippings and photographs. In total there were ten missions between 1968 and 1975. During this period, the team conducted detailed surveys of the seamount; installed, tested, and recovered scientific instrumentation and navy equipment (explosive anchors and sonar instruments); and conducted diver physiology studies. A submersible, the *Sea Otter*, was used in survey work during the summer of 1972.

Most of the instrumentation installed on Cobb over the course of the project was dedicated to biological and physical oceanographic research.[131] The dive team would wake at five thirty, get their dive systems set up before breakfast, and after predive briefings, prepare the Zodiac and scuba gear. The first dive might begin at eight with diving operations continuing through the day until about six. After that the divers were kept busy with maintenance chores and record keeping until nine, when they could finally retire to their quarters.

During the first mission to Cobb in the summer of 1968, it became clear that funding would be insufficient. The diving accident that forced the first mission to be aborted raised safety concerns with organizers and funders. A federal austerity program cut available funds just as the project was starting, requiring reliance on the collaborating institutions. Nevertheless, for ten years the program carried on, cobbling together funding year by year. In the lead-up to the 1969 mission, in order to generate publicity and financing, the team modified a navy-surplus net buoy to serve as a small underwater habitat in Puget Sound, playfully named "Man-in-the-Sound." A popular fishing wharf near Edmonds, Washington, was selected as the installation site. Over a hundred divers visited the two-person habitat during its three-day deployment in November 1969, during which they gathered biological samples for the local community college. A sounding device dropped by airplane was recovered at night in a proof-of-concept test for the delivery of supplies to the planned underwater habitat on Cobb. The success of this smaller operation gave a lifeline to the ambitious seamount program.[132] In May 1969, in the program's second summer, the aquanaut Robert Sheats, veteran of the navy's Sealab program, served as diving supervisor.[133] Campbell took on the role of his assistant, and subsequent success of the 1969 mission reassured organizers that the dives could be done without undue risk.

Sea Use aquanauts helped lay down and secure a five-and-half-kilometer "chain highway" running east to west across the summit. Reference markings were laid along every 30 meters of this highway, including one street sign midway jokingly labeled "Aquanaut street" and "Cactus Ave." On the reverse it read "Seattle 300 miles" and "Japan 6000 miles."[134] Despite these achievements, however, plans for establishing an underwater habitat were abandoned.[135]

As budgets shrank and federal interest in aquanaut programs waned, the Sea Use program devolved into a regional effort, covered in Washington news media but largely ignored elsewhere—even though it has the distinction of being the largest open-water scuba survey project ever conducted. The scientific data it produced was valuable, providing a better understanding of seamount ecosystems and geology. And work on Cobb helped train a generation of regional divers who went on to work in the commercial diving industry that burgeoned with the expansion of offshore oil exploration. The only remaining physical vestige of Sea Use today is the long chain highway draping Cobb's summit, fouled by trawling nets, and slowly being reclaimed by the original denizens of the mountain.

Conclusion

Outer space, not inner space, became the final frontier to capture late twentieth-century popular imagination. When President John F. Kennedy spoke of the race to the stars, it was, somewhat paradoxically, in ocean metaphors that he described this unfamiliar terrain.[136] "Only if the United States occupies a position of pre-eminence can we help decide whether this new ocean will be a sea of peace or a new terrifying theater of war," he declared in his famous 1962 "we choose to go to the moon" speech.[137] But even though Cold War political rhetoric focused on the space race, military games of cat and mouse between the Americans and the Soviets played out in the oceans. In these submerged battlefields, the strategic advantage belonged to those who could use the terrain to their advantage. As the American oceanographer and naval officer Robert Ballard later recalled, "During the Cold War, we tried to box up the Russians, whether it was along the Greenland-Iceland gap or the entrance to the Bosporus, or the entrance to Gibraltar. The idea was chokepoints ... where you're in the terrain and [the enemy is] really silhouetted above your head."[138] With submarine warfare, and such strategic thinking, the deep sea retained its military importance.

By 1970, the end of the underwater habitat craze was clearly in sight. Writing in *New Scientist*, a senior researcher at the British National Institute of Oceanology lamented that although some had predicted that "soon the silent darkness of the ocean would glitter with the lights of underwater cities, and throb with the noise of underwater motors, pumps, power stations, mines, fish factories, chemical plants, and transport systems, while men flitted from structure to structure like fish themselves," there was "as yet no sign of this happening." He dismissed the augury as misguided "romanticism" and, noting the dangers and cost of saturation diving, accurately predicted that "competitive unmanned systems" would take precedence over underwater human labor.[139] He also accurately predicted that ship-borne decompression chambers with pressurized transfer capsules for reaching depth would be cheaper and safer than underwater habitats. Already by 1967, American oil companies began funding experiments with saturation diving transfer capsules. Ocean Systems and Esso carried out a research program with their Advanced Diving System (ADS III) and a deck-mounted decompression chamber in the Gulf of Mexico in 1967. At a depth of 194 meters divers carried out simulated work on an oil well head, transferring several times from the surface to depth to test the safety of the deck-mounted pressure chamber.[140] For commercial purposes, underwater habitats had become obsolete.

The American program for undersea manned exploration was not the only program to slow by the late 1960s. By 1970, owing to a lack of public support and budgetary restrictions imposed by the Vietnam War, NASA's Apollo program began winding down.[141] In 1978, NOAA administrators proposed the creation of Oceanlab, an undersea mobile habitat to go into operation by 1982. Steven Anastasion, director of NOAA's Ocean Engineering Office, described Oceanlab as a "national facility" to be "operated by NOAA for cooperative use by all federal agencies and by universities and oceanographic institutions." Unlike previous underwater habitats, Oceanlab would be able to operate "under varying weather and sea conditions, particularly those of the northern latitudes where our oceanic interests [were] being increasingly focused."[142] The project, however, was shelved. At an estimated cost of $21.5 million, it would have been the most advanced facility of its type in the world.[143]

It is tempting to conclude that the story related here is one of a technological dead end. Clearly, the continental shelf has not been colonized, and today's oceans are still more overfished than carefully managed. Nevertheless, the world we live in today is very much a product of the man-in-the-sea era. Underwater engineering, pioneered in the 1960s and 1970s, benefits the offshore oil industry. Underwater medicine is used by commercial divers. With the help of private investment, oceanic exploration is experiencing a rebirth. Fantasy and fear continue to motivate action—with real consequences.

From the point of view of aquanauts of the 1960s and early 1970s, colonization of the seas was conceived as part of a larger imagination of an imminent human future. Past failures often reveal possibilities that were once thought viable even if never realized. When trying to understand why the underwater habitat-building period ended, it is fair to argue that it came to a close primarily because of shifts in federal policy and as a result of inadequate funding, rather than because of insurmountable technological problems. The mechanical difficulties of living on the seafloor were challenging but not impossible to surmount—travel to the moon and the establishment of orbiting space stations were arguably far more difficult to achieve.[144]

In August 2007, two Russian minisubmarines touched down on the seabed on the Lomonosov Ridge in the Arctic Ocean, planted a titanium

Russian flag, and symbolically laid claim to several billion dollars' worth of oil and gas.[145] Three years later, Chinese scientists descended in the research submersible *Jiaolong* to plant China's flag on the floor of the South China Sea.[146] Early in the twenty-first century, China aimed to bolster marine territorial claims by building artificial islands and infrastructure on top of submerged coral reefs. The country sought to justify these actions to the international community as being aids to marine science, fisheries research, and navigational safety. China announced plans in 2016 to build an underwater "manned deep-sea platform" to be operated in the South China Sea. This oceanic "space station," according to a Chinese government press release, would host a dozen crew members for up to a month. As of 2019, no blueprints or cost estimates have been released.[147] Russia announced in 2017 similarly ambitious plans for underwater construction of Project Iceberg in the Arctic.[148] The story of Project Sea Use provides a precedent of territorial dilemmas and debates that continue today, unresolved. Oceanographic science, based in the field and requiring the installation of infrastructure beyond national jurisdictions, will not be able to avoid becoming embroiled in them.[149]

As we see in Chapter 5, by the late twentieth century some scientists and policy makers, beginning to recognize the fragility of marine ecosystems, were led to reimagine the future of the ocean once more. No longer simply a territory to colonize, they reconceived it as space where a more harmonious relationship not only among nations but also between humans and the natural world might flourish.

5

Ocean Science and Governance in the Anthropocene

> The Oceans are our great laboratory for the making of a
> new international order, based on new forms of international
> cooperation and organization, on new economic theory,
> on a new philosophy or *weltanschauung*.
>
> Elisabeth Mann Borgese, The Future of the Oceans

> This is the first age that's ever paid much attention to the future,
> which is a little ironic since we may not have one.
>
> quotation often attributed to Arthur C. Clarke

> So have I seen Passion and Vanity stamping the living
> magnanimous earth, but the earth did not alter her tides
> and her seasons for that.
>
> Herman Melville, Moby Dick

It is the year 2000. The coastal powers have extended their sovereignty to the centers of the oceans. Cargo and military vessels must pay tribute as they pass from one sovereignty zone to another or as they transit straits through which passage once was free. Conflict between the "have" and "have-not" countries, as governments jostle over the resources of the seabed, keeps the world in a state of tension. Fish are a rarity; the few species that survive taste rather odd, for they inhabit an element befouled by enormous amounts of pollution. In most coastal areas, swimming in the sea is forbidden by law.... The environment needed to sustain life on earth is wearing away.[1]

This was the grim prediction for the twenty-first century made by Richard A. Frank (1936–2014) in the *New York Times* in the spring of 1975. Two years later, President Jimmy Carter nominated Frank to serve as administrator of the National Oceanic and Atmospheric Administration, a post he held until 1981. As the second decade of the twenty-first century draws to a close it is clear that some elements of Frank's nightmarish prediction have come to pass. The South China Sea has become a simmering pot for potential future geopolitical conflict, and there are signs that the retreating sea ice in the Arctic has set off a race between polar nations to claim the mineral and petroleum riches below the seabed. In general, the detrimental effects of the Anthropocene on the marine environment are staggering. Between 1950 and 1989, the open-ocean industrial fishing catch increased by a factor of more than ten.[2] A 2016 assessment by the United Nations reports that 89.5 percent of global fish stocks are either fully fished or overfished, with projected commercial fishing increases of 17 percent by 2025.[3] And as historians of marine ecology have shown, the species of fish harvested today are different from those our ancestors caught. Not only have we begun to hunt new species as previously harvested commercial stocks have collapsed; the fish too have changed. In response to the selective pressures of human exploitation, the average size of many commercial species has shrunk as mature and large fish disappear.[4] In addition, we have long used the oceans for our collective trash disposal. It is estimated that only 1.6 percent of the oceans have been placed under protective jurisdiction—so-called marine protected areas. By some estimates, the total mass of plastics in the ocean will exceed that of fish by 2050; some eight million tons of plastic waste enters the oceans each year.[5] A 2009 study found that only 13.2 percent of the global oceans can still be considered pristine—not affected in major ways by fishing and pollution. Daniel Pauly, the fisheries scientist who first coined the term "shifting baselines" expresses the situation succinctly; in his view, we are witnessing an "aquacalypse."[6]

One of my early childhood memories is of the Spanish fishing vessel *Estai* being escorted into St. John's harbor in Newfoundland by the Canadian coast guard. The *Estai* was seized in March 1995 for fishing turbot just beyond Canada's territorial exclusive economic zone, an incident that led to the turbot war between Canada and Spain. Newfoundland fishermen were still smarting from the collapse and closure of the cod fishery, and the Canadian government feared that turbot (also known as Greenland halibut) would suffer the same fate as the cod. The Canadians suggested a quota limit, which the European Union rejected, resulting in the crisis. Eventually, the matter was settled in the International Court of Justice. I was too young to understand the details of the conflict. What I do remember is the anger of the Newfoundlanders and the way they jeered the Spanish crew being led off the ship. Newfoundlanders felt that Spaniards had stolen something that belonged to them and that there weren't enough fish for everyone—never mind that the Spanish were fishing beyond Canadian territorial waters and had fished the Grand Banks for centuries. The conflict was preceded, and undoubtedly shaped, by the collapse and closure of the Newfoundland cod fishery three years earlier, an event that one Canadian government report described in apocalyptic terms: "[It is] a calamity which threatens the existence of many of these communities throughout Canada's Atlantic coast, and the collapse of a whole society. . . . We are dealing here with a famine of biblical scale—a great destruction."[7] Certainly for Newfoundlanders, a way life that had organized and given meaning to families and communities for many generations had come to an abrupt end.

In the Outer Battery neighborhood of St. John's in which I grew up, the fishermen always repaired their wharfs in the spring after the damage wrought by winter storms. With the fishery gone and no more reason for the hard labor of maintaining the infrastructure that made the small-craft inshore fishery possible, the sea gradually reclaimed part of the shoreline. Every year I watched as, piece by piece, the vestiges of that ancient trade were swallowed up in the roiling surf and swept out to sea. It gave me the impression of an ocean pillaged and barren. In reality, the despoliation of the oceans began decades earlier, although it accelerated in the second half of the twentieth century. For the tipping point we might point to the launch in 1954 of the British ship *Fairtry*, the first factory-freezer trawler in the world, processing up to six hundred tons of fish a day and allowing the harvest of ocean-caught fish on a truly massive scale.[8] A 2018 estimate is that industrial fishing occurs in more than 55 percent of ocean areas—a planetary surface four times the size of that used for agriculture.[9] Tragically, contemporary industrial fishing is also enabled by government subsidies (estimated at $20 billion per year).[10] Without these subsidies, which allow vessels to travel farther and fish for longer periods at greater capacity, as much as 54 percent of high-seas fishing grounds would be unprofitable.[11]

It is difficult not to be pessimistic when we consider the likely fate of the oceans. But we must also strive to be conscious of how such pessimism could be shaping contemporary science and marine policy. A first step is to determine when misplaced confidence in increasing human ability to control the marine environment was replaced by fears of a global systems collapse. As discussed in Chapter 4, Cold War fears, similarly to the threat of war in the first half the twentieth century, reframed ocean spaces as zones of conflict and strategic military concealment. Public optimism in the United States in the victorious aftermath of World War II and the successful messaging campaign of President Dwight Eisenhower's Atoms for Peace program also birthed techno-utopian dreams of underwater cities and nonviolent uses of atomic power. But by the 1970s, as the proxy wars of the Cold War took their toll and the horrors of nuclear weapons became more widely understood, this dream had collapsed.

In 1974, the *Mutsu*, Japan's first nuclear-powered cargo ship, leaked radioactive waste eight hundred kilometers off the coast of Japan during its maiden voyage. In an effort to stem the leak, the crew stuffed boiled rice and hundreds of socks into the reactor's pressure chamber. The drama, which unfolded in the global media, added fodder to the arguments of the growing antinuclear movement. Attaching their cause to the publicity generated by the *Mutsu* incident, one activist group, Business and Professional People for the Public Interest, ran ads in college newspapers asking students to send their old socks to Dixy Lee Ray, chair of the Atomic Energy Commission, as a protest against the dangers of nuclear energy.[12] In the words of one observer, the incident "probably did more to set back the cause of nuclear marine propulsion than anything in the industry's worst nightmares."[13] The failure of the *Mutsu* can be seen as the death knell to the dream of peaceful uses of atomic energy at sea.

In this chapter I again examine debates surrounding the use of marine spaces in the twentieth and early twenty-first centuries, but I now focus on changing perceptions of the marine environment. Specifically, I describe the transition when oceans were reconceived from an environment that could be conquered and controlled to something fragile, broken, and dying. Examining this transition will also require exploration of how modern technology has reshaped marine science and geopolitics in the early twenty-first century. But before looking at these developments, let us first examine the work of two twentieth-century figures whose work falls on either side of the divide outlined here. The first is the American engineer Carroll Livingston Riker (1854–1931), and the other is Elisabeth Mann Borgese (1918–2002), daughter of the Nobel Prize–winning author Thomas Mann. Temporally at opposite ends of the twentieth century, Riker and Borgese imagined the ocean environment quite differently. Riker regarded the ocean as a simple system that could be harnessed and made to serve humanity, whereas Borgese depicted the oceans as complex, fragile, and under threat by humans. Of the two, Borgese was by far the more successful in bringing her plans to fruition, but each imagined oceans as terrain demanding a new international political regime.

"Mr. Riker's Amazing Plan"

One of the most formidable natural phenomena on our planet, the Gulf Stream has long attracted the interest of marine scientists. Its power to shape the global climate has fascinated naturalists ever since Benjamin Franklin first mapped it in crude outline. "The Gulf Stream is a historical agent," wrote Prince Albert I of Monaco in 1886.[14] Prince Albert devoted considerable energy and resources to mapping its vagaries during his oceanographic expeditions in the North Atlantic. He was followed by numerous other marine scientists who devoted their lives to studying this current, believed to hold the key to understanding global ocean circulation.

Matthew Maury, the American naval officer and oceanographer, wrote in 1855, "There is a river in the ocean. In the severest droughts it never fails, and in the mightiest floods it never overflows. . . . It is the Gulf Stream. There is in the world no other such majestic flow of waters. Its current is more rapid than the Mississippi or the Amazon."[15] Although later researchers determined that the Gulf Stream is better understood as a series of linked turbulent eddies rather than as a river, they continued to describe this distinct marine feature in similarly adulatory terms. For the twentieth-century oceanographer Henry Stommel, who made major contributions to the study of its physical properties, it was "a grand natural phenomenon."[16]

Carroll Livingston Riker became fascinated with the Gulf Stream in the early twentieth century (see Figure 5.1). A mechanical engineer, he had some notable achievements to his name. He had built the first refriger-

THIS MAN PROPOSES TO CONTROL THE HOURLY FLOW OF 90,000,-000,000 TONS OF WATER.
He is Carroll Livingston Riker, a successful American hydrographic engineer.

Figure 5.1 Carroll Livingston Riker. Robert G. Skerrett, "Warming the North Atlantic Coast," *Technical World Magazine*, March 1913, 62.

ated warehouse in the world, built the first cold storage system for a transatlantic steamer, and in 1887 won a government contract to fill in the Potomac flats near Washington, D.C., using a pumping dredge of his own design. By the end of his life he held more than twenty patents. He was an adept promoter, but his aspirations frequently overstepped his ability to achieve his projects.

In 1928, Riker built a room-sized scale model dam and canal system, complete with running water, valleys, and levees and installed it in the basement of the Capitol in Washington, D.C., to demonstrate his plan for controlling the flooding of the Mississippi delta. His plan, described in an interview in the *New York Times*, was soundly mocked by an editorial in *Engineering News*, "Amateur Engineering in the Newspapers," and Riker was reprimanded for not soliciting the opinion of engineering societies before speaking to the press. But even the ambition of this project paled in comparison to another, which he had promoted in an earlier self-published pamphlet.

For a mere $190,000,000 (a bargain, he argued, if compared to the cost of the Panama Canal), a three-hundred-and-twenty-kilometer-long jetty could be extended out onto the Grand Banks from the island of Newfoundland to redirect the course of the Labrador Current. The Gulf Stream would then be freed to move farther north into the Arctic Ocean, warming the coasts of Newfoundland and eastern Canada, opening ice-bound harbors for Atlantic shipping, and diminishing the danger of icebergs—like the one that had recently sunk the *Titanic*. Riker went so far as to predict that the resultant melting of the polar ice cap would affect the balance of the earth in its solar orbit, bringing heat and light to the Arctic and producing "long twilights north of New York, and almost continuous day in Scotland for considerable periods, without any periods of continued night."[17] Riker proclaimed in a subheading on his pamphlet's title page, "MAN CAN CONTROL ALL."

This faith attracted some enthusiastic believers. A writer for *Technical World Magazine* enthused, "There was a time when a project of this sort would have met with ridicule—the hydrographic engineer had not then accomplished the marvels which have since become commonplace by repetition—but today both mechanical facilities and practical experience have altered the viewpoint." In the writer's view, only the need to secure "international agreement" stood in the way of the project's rapid implementation.[18] Another writer expressed some skepticism at this "amazing plan," but conceded that the proposal was "made by Mr. Riker in all seriousness and sincerity."[19] What Riker knew of the Gulf Stream he had gleaned from reading the work of some of the leading marine scientists of his day.[20] "Nature seems to favor this undertaking in nearly every way," Riker proclaimed, noting that in his estimation a jetty extending out onto the Grand Banks merely restored a previously existing coastline. It is tempting to dismiss Riker's project as an impossible utopian fantasy. But, as the historian James Fleming points out, Riker lived in a period of astonishing feats of engineering, foremost among them the building of the Panama Canal, completed in 1914. Riker gained the support of Congressman William Musgrave Calder from New York, who proposed forming a Commission on the Labrador Current and Gulf Stream.[21]

In addition to harnessing the forces of the ocean, Riker dreamed of establishing a new political geography on the seas. In a 1915 pamphlet, wordily titled *International Police of the Seas: A Simple, Feasible, Common Sense Plan to Bring About Lasting Peace*, Riker proposed that the only way to bring an end to naval rivalry between Britain and Germany was to form a "neutral representative government for the seas." This body, composed of international representatives, having its own constitution, and "preferably" headquartered on a small island in the North Atlantic, would have the power to purchase naval vessels and islands for naval bases. Such military power would allow it to function as an international police, to "render service to vessels in distress, and to forcibly prevent any fighting on, over, or under the seas." With a conviction certain to be heatedly opposed by nationalists, Riker also proposed that an "International Police of the Seas" should be given jurisdiction to maintain the neutrality of all major ship passages, including the Panama Canal, the Suez Canal, and the Strait of Gibraltar.

Riker asserted that his views were based entirely "upon mechanical facts, and upon deductions emanating from . . . experience in submarine engineering." He had not himself designed submarines, but he had constructed torpedoes during the Spanish-American War, and in his view the submarine would eventually make battleships obsolete. Riker also offered a characteristically dramatic prediction for the future of submersibles.

"It will be found no dream," he wrote, "but mechanically practical and easy, within 6 months to build 50 slow freighters carrying 5000 tons each, submersible at will that could circumnavigate the globe without communicating with any base for supplies." By these means, "the future control of any part of the seas by any nation" would become "impossible."[22] Like many of Riker's other schemes, this one never came close to realization, despite the support of Congressman Calder.

By the end of his life, Riker had become a fixture on Capitol Hill, often seen explaining to passersby his model of the Mississippi flood control system set up in the basement of the Senate office. A journalist described him as a "white-haired, browned but fragile gentleman" who loved to denounce the Army Corps of Engineers "in masterpieces of invective fit to delight any lover of words." As his health began to fail, Riker retreated to a house in Virginia that he had designed by himself and set up in the air on a tree stump. It was here that he died at the age of seventy-eight in 1931, his grandest schemes unrealized.

Nevertheless, fantasies concerning human manipulation of the Gulf Stream did not die with him. As late as 1949, some politicians in Britain expressed fears that the United States might soon gain the power to "turn off the heat" in Europe. Lord Strabolgi, a member of the House of Lords and a former navy officer, was quoted in American newspapers voicing concerns that "there are hotheads in the United States who talk quite glibly—'in the event of your going communist,' as they call it—of diverting the Gulf Stream by the use of immense atom power in the Atlantic." This claim was met with mocking incredulity by American scientists, one of whom responded, "Any claim that the course of the Gulf Stream can be substantially altered by any known man-made means is, in my opinion, fantastic."[23]

Despite being the subject of scientific observation for over two hundred years, many aspects of the Gulf Stream remain poorly understood even today.[24] The Gulf Stream is a highly dynamic fluid system, more akin to a shifting river delta than a fixed mountain range. Sailors crossing it rely on eddy position maps that remain reliable for only a few days. One of the challenges of studying global ocean circulation is the need to observe complex fluid dynamic systems over enormous ocean-scale spaces. It was only in the second half of the twentieth century, as humans inadvertently released into the marine environment chemicals that occur rarely in nature—radioactive tritium from atomic testing and chlorofluorocarbons from refrigerants—that a suitable means was found for tracking surface waters.[25] Physical oceanographers have since developed laboratory experiments that reproduce ocean circulation at a small scale or through computer modeling. And now supercomputers model ocean circulation and visualize eddy formation on a global scale.[26] Humans may, in fact, have begun to alter the flow of the Gulf Stream, albeit unintentionally. As global temperatures continue to rise because of anthropogenic climate change, some scientists are voicing concern over the long-term stability of ocean circulation in the Atlantic, which they fear can be disrupted by an influx of fresh water from melting glaciers.[27] In fact, circulation in the Atlantic slowed in the 2010s.[28] Some hypothesize that the observed slowdown may indicate a longer-term naturally occurring oscillation.[29] These studies reveal the limitations of our knowledge, as well as the frightening, accelerating impact human activities have had on the planet since the beginning of the Industrial Revolution. It appears that although humans have power to change the oceans in many ways, they are not in control. Only since the second half of the twentieth century has this reality received broad, authoritative recognition.

"The Mother of the Oceans": Elisabeth Mann Borgese

Elisabeth Mann Borgese (1918–2002) was one of the twentieth century's leading scholars and a prolific author on ocean governance. She argued that the ocean demanded an alternative model of international political order—a counter to the established regime that hinges on the existence of nation-states. Borrowing a term from Mohandas Gandhi, she described the new global order she called for as the "oceanic circle."[30] Unlike Riker, Borgese did not seek to control the marine environment but to protect it, and all of nature, from human depredation. She held the conviction that restoring balance with the natural world at the local level could in turn foster a regime of peaceful international relations at the global level. Colleagues, dazzled by her tireless advocacy for marine protection, referred to her with reverence as "the mother of the oceans."[31]

Borgese was born in Munich, Germany, in 1918, but in 1933 moved with her family to Kuesnacht, near Zurich, Switzerland. Her father, the celebrated Nobel laureate Thomas Mann, went into exile when the Nazis came to power. Elisabeth recalled a repressive childhood, her early interest in music discouraged; "I was disturbed that I was a girl," she later told a reporter, "it was a handicap reinforced by both parents, who were male chauvinists."[32] Nevertheless, Elisabeth attributed to her father her early introduction to the importance of the ocean. "My father was proud to show us the sea," she wrote. But what most impressed her was the sight of the horizon. "What is after the horizon?" she recalled asking her father, to which he responded, "The horizon ... and after, ... again the horizon." The concept of endless horizons offered both philosophical and aesthetic appeal to young Elisabeth, who compared the ocean horizon, "where finitude becomes infinitude," to the horizons of the universe. Love of the sea, she later wrote, became the "emotional and aesthetic" foundation for her life.[33]

In 1939, the Mann family moved to the United States, and Elisabeth married Giuseppe A. Borgese, thirty-six years her senior, a professor of Italian literature at the University of Chicago. In Chicago she cultivated an interest in international law. In the aftermath of World War II, as part of the World Federalist Movement, a group of University of Chicago academics, including Elisabeth, her husband, and former university president Robert M. Hutchins, convened as the Chicago Committee to Frame a World Constitution. The group produced a monthly magazine, *Common Cause*, and in 1948 the University of Chicago Press published their *Preliminary Draft of a World Constitution*. They advocated the dissolution of nation-states, claiming that they were "by definition and nature the enemy and antagonist of the World State."[34] For her own part, Borgese was skeptical that the United Nations could ever achieve global peace. In her estimation, the UN "represented nothing else but the powers victorious in the Second World War who had made up their minds that their task was to avoid a new war."[35] When word of the Chicago Committee got out, the *Chicago Tribune* gave the story first-page billing and warned of the "supersecret" constitution's "alien," "socialist," and "Marxian" principles.[36] Others lambasted the proposal as "fantastically utopian, unintelligent, and naïve."[37] The project was nevertheless intellectually formative for

Borgese, later remembered as saying, "The Utopians of today are the realists of tomorrow; the realists of today are dead tomorrow."[38]

After her husband died in 1952, Borgese moved with her three children to Florence, Italy, and subsequently, in 1964, to Santa Barbara, California. Although she had no formal graduate training, she was a prolific scholar, editing a journal for the Ford Foundation and publishing works of fiction and a treatise titled *Ascent of Woman* in 1963. She also began her own experiments in animal intelligence, publishing a book on the topic, *The White Snake,* in 1966. In Santa Barbara she took a position as a senior fellow in the Center for the Study of Democratic Institutions (CSDI), a policy think tank founded by her Chicago friend Robert Hutchins in 1959.[39] The CSDI, which was largely privately funded, especially by the Ford Foundation, organized speaker series and conferences, including one series titled Pacem in Terris.[40] Borgese's work at the CSDI dealt with human rights and disarmament. Like the Chicago Committee, the CSDI was denounced by the Right for its liberal agenda of the Left.

Through her involvement with the CSDI, Borgese also became involved with the Club of Rome—founded in 1968 by a group of high-profile businessmen, politicians, and scientists—joining as the first female member in 1970. The Club of Rome, which continues to operate, is described by its members as "an organization of individuals who share a common concern for the future of humanity." The club published a report in March 1972, *The Limits to Growth,* that became an immediate international bestseller.[41] The report summarized the results of a study carried out by researchers at MIT who used a computer model called World3 to project the interactions of population, pollution, food production, industrial production, and consumption of nonrenewable natural resources between 1900 and 2100. They concluded that continued growth of the global economy would confront planetary limits in the twenty-first century, resulting in the collapse of the global population and the world economy.[42] *The Limits to Growth* added to a growing body of scholarship, begun in the 1960s, that warned of increasing environmental destruction: Rachel Carson's *Silent Spring,* published in 1962; Paul Ehrlich's *The Population Bomb,* in 1968; and Barry Commoner's *The Closing Circle,* in 1971, all became best sellers. This period also saw a series of oil spill disasters that drew public attention to problems of marine pollution: the 1967 shipwreck of the supertanker *Torrey Canyon* released 120,000 tons of crude oil that washed up on the coasts of England and France, and in 1969 an oil platform off the coast of Southern California blew out, releasing 760 tons of crude oil and coating recreational beaches.[43] Both incidents were the subject of book-length exposés describing the toll on wildlife and failures of clean-up efforts.[44]

With growing public concern about the impact of humanity on the planet, and on oceans in particular, the environmental movement and early efforts toward marine environmental protection were born. The first marine park—the John Pennekamp Coral Reef State Park in Key Largo, Florida—was established in 1963, and the Great Barrier Reef gained protection in the mid-1970s. At the same time a public aware of how little was known about the ocean was becoming attuned to increasing revelations of the complexity of marine life. The 1963 film *Flipper*, detailing the adventures of a boy and his dolphin friend, was immensely successful.[45] And when biologist Roger Payne released a recording of humpback whale songs in 1970, the LP record sold over two million copies. Jacques Cousteau's film *The Silent World* presented coral reefs as complex hosts to multivarious life forms and as crucial to human well-being. Antiwar and environmentalist groups found early success, such as the Greenpeace Save the Whales campaign.[46] The influence of the Club of Rome and the environmental movement on Borgese's work is evident in her later approach to ocean governance—interpreted as work to find solutions to a series of interrelated problems considered as a whole.[47]

Borgese organized her own conference, Pacem in Maribus, in Malta in 1970. She was inspired by a four-hour passionate speech to the UN General Assembly delivered in 1967 by Arvid Pardo, the UN ambassador of newly independent Malta, who urged global powers to recognize the importance to all humanity of the world's oceans and to declare the seabed and ocean floor lying beyond national jurisdiction as part of the common heritage of humankind. Pardo's speech was instrumental in bringing about negotiations that led to the 1982 UN Convention on the Law of the Sea (UNCLOS III).[48] In his speech, he argued that humanity had arrived at a crossroads. As he explained in his opening words, "Man, the present dominator of the emerged earth, is now returning to the ocean depths. His penetration of the deep could mark the beginning of the end for man, and indeed for life as we know it on this earth: it could also be a unique opportunity to lay solid foundations for a peaceful and increasingly prosperous future for all peoples."[49] Finding inspiration in Pardo's words (Arvid Pardo became a lifelong friend and collaborator), Borgese began to see the oceans as the ultimate political laboratory for a new world order (see Figure 5.2).[50] The Pacem in Maribus conference and subsequent meetings culminated in the 1972 creation of the International Ocean Institute headquartered in Malta, a nonprofit, nongovernmental organization with a mandate to foster improved approaches to ocean management and governance.

Borgese never formally studied maritime law (although she later taught political science at Dalhousie University in Canada), nor was she a marine scientist. Yet she had an enormous impact on the development of

Figure 5.2 Elisabeth Mann Borgese and Arvid Pardo.
Courtesy of the Elisabeth Mann Borgese Fonds
(MS-2-744), Dalhousie University Archives, Halifax,
Nova Scotia.

marine policy and ocean governance, remaining active up until her death in 2002 at age eighty-three. In an essay written in 1983 she rhetorically asks, "Why on earth am I doing what I'm doing?"[51] In answer, she cites the long cultural interplay between humans and the sea, recalls the love of the sea instilled in her by her father, and refers to Alister Hardy's aquatic ape hypothesis (see Chapter 4). Borgese suggested that the oceans have "played an enormous role in the evolution of human life and culture, in the dreams of men" and that this influence was "bound to grow." Citing her father's description of the ocean as "ultimate and savage, [of] extra-human magnificence," she points out the irony, and apparent blasphemy, of seeking to impose a new legal order on the oceans. Yet, she argues, "our humanizing efforts, our utilitarianism does not detract one whit from the concept of the enormity and wildness of the sea. It is not the oceans we want to dominate and regulate, it is human activities and human encroachment."[52] Stating the pressing need to control human powers over nature, Borgese also acknowledged the arrival of a "humanized ocean" that will come to redefine humanity's relationship with nature.

> Based on the dream of common heritage, we see, emerging from the sea, a new "ecological consciousness," a different vision ... of man's relationship to nature in general and to the sea in particular. We see a vision of human evolution and history, not as a confrontation with nature but as part of nature; not called by any god to subdue her, but led, by nature herself to co-operate.... For the environment in general (not only the sea), both natural and social, is an extended mirror of man's soul. For better or worse, just as we perceive ourselves, so we see the world around, oceans and all.[53]

It is difficult to imagine that Carroll Livingstone Riker, or even the environmentally conscious legal theorists of the 1960s, could have fully anticipated the scale and impact of anthropogenic climate change in the twenty-first century. As we have repeatedly witnessed, predictions for the future routinely fall short. However, from today's vantage it is nevertheless possible to point to some of the changes already affecting the oceans of the Anthropocene. These developments may give us vague glimpses of an otherwise still-elusive ocean of the future; but more importantly, they attest to some of the ways oceans of the future are being imagined and created today.

Probing Anthropocene Oceans

For most of human history the ocean has remained a remotely sensed terrain, vast swathes of which remained unseen and unknown. As oceanographers like to observe, we have better maps of the surface of the moon and Mars than we do of the seafloor. This was made abundantly clear by the disappearance of Malaysian Air flight 370 in 2014. What is known about the geography of the ocean has long been limited by what Anne-Flore Laloë describes as a "shipped perspective"—a viewpoint constrained by the limited space aboard a ship. The lines on charts showing the track of a vessel at sea are best understood as representative of a "temporal relationship with the ocean, not a spatial one."[54] Hampered by limitations of the ship, oceanographers have long had to wrestle with the extreme conditions of work at sea. As the oceanographer William A. Nierenberg once said, "To oceanographers the sea is an enormous and restless antagonist. The work is nowhere near as glamorous as it's supposed to be—it's tough, rough, very difficult."[55] Another scientist, the marine geologist H. William Menard, has described the ocean simply as "regrettably unstable" (see Figure 5.3).[56]

Modern research vessels are much more technologically advanced than those that first set out in the nineteenth century. Yet the "shipped perspective" of the late nineteenth century and much of the twentieth century resulted in "undersampling" and subsequent misinterpretation of the data collected.[57] Although research vessels made consistent measurements at different "stations" as the ship crossed the ocean, differences between stations were interpreted as variations in space rather than in time; and observations at any one station could not be repeated, leading to an underestimation of the importance of seasonal changes in the ocean basin.

Oceanography's gradual transformation into a global systems science mirrors broader social, political, and economic concerns with global climate change. In turn, concern about global marine degradation has rekindled scientific interest in international scientific collaboration. For

Figure 5.3 Deploying a carousel for water sampling in heavy seas. Photo by Craig Dickson, © Woods Hole Oceanographic Institution.

instance, citing marine pollution as a "clearly international" problem, the oceanographers Henry Stommel and E. D. Goldberg appealed in a 1969 editorial in *Science* for "an international laboratory, equipped on a scale commensurate to global problems." Projects had to be scaled up, they argued, and they proposed expeditions for the following decade that would require "twenty-five hydrographic survey ships, forty oceanographic and fishery research vessels, ten fishing vessels, for systematic resource surveys, eight weather ships, an aircraft carrier with two escort ships, three deep-sea drilling vessels and five drilling barges, as well as some submarines including one nuclear power."[58] But Stommel also insisted that oceanography, faced with the urgency of climate change, requires a "sharp focus for an enhanced emergency research effort," which he described as comparable to the procedures of wartime science.[59] By the late twentieth century, internationalism was thus being reframed as a buttress against a new threat: global climate change.

Only with the advent of satellite technology did a new era of constant sampling begin. Stable instrument platforms, unlike a moving vessel, allow sampling in an identical space over extended periods. These developments in oceanographic technology reflect an enduring effort: to extend the reach of marine science in the field while shortening the time between data collection and analysis. The oceanographer Robert Ballard, the foremost proponent of telepresence, has described the method as effectively endowing the human body with cybernetic enhancement, achieving nothing less than species modification: "Now we can cut the tether—the one that binds our questioning intellect to vulnerable human flesh. Through telepresence, a mind detaches itself from the body's restrictions and enters the abyss with ease, and with lightning-quick fiber optic nerves."[60] The expansion of human sensory immersion has been accompanied by a reduction in the time needed for data collection.[61] In fact, there is now such an abundance of data flowing ashore that there is far more than can be analyzed by unaided human workers; one computer scientist describes the flow of oceanographic data now coming ashore as a "deluge."[62] Scientists hope that advances in artificial intelligence and machine learning will eventually make it possible to transfer more fieldwork decision-making to machines.[63] Already algorithmic and bioinformatic analysis is regularly used to sort and interpret oceanographic and biological data logs.

Thus, we are again at the cusp of a technological revolution that promises to bring us even closer to a panoptic view of the ocean. Using satellites, autonomous vehicles, and remote sensing, the seafloor will soon be mapped in unprecedented detail. But the efforts of oceanographers to develop sophisticated remote-sensing technology are rapidly being matched by other interests seeking enhanced access to the remote regions of the ocean. The offshore oil industry, having profited from marine engineering advances powered by the Cold War, has dramatically augmented the capabilities of submarine robotics, remote sensing, and bathymetric mapping for industrial applications.

These technological advances once again push us to look to the future to imagine the opportunities—and pitfalls—these tools will bring. On the one hand, some have called for scientific collaboration with industrial partners, in the hope that commercial technology might be adapted to answer biological questions.⁶⁴ On the other, some marine conservationists fear that technological developments will permit industrial exploitation of unparalleled scale, outpacing scientific observation and analysis. One review article from the BBC summarized the alternative futures that might be favored by a single scientific advance: "A global bathymetric map ... would certainly offer a better understanding of our blue planet, but it also might plunge us into a realm once reserved for science fiction: robot submarines, underwater volcanoes, sea jewels, coral with pharmaceutical properties, Wild West maritime law, toxic sediment plumes, and an ocean-based enterprise curiously devoid of humans or ships. Once the map is made, will it be used as a tool for responsible management and conservation, or wielded like pirate's bounty, a guidebook to extraction and exploitation?"⁶⁵ Manganese nodules, first discovered on the seafloor during the *Challenger* expedition in the late nineteenth century, have long tempted prospectors seeking nickel, copper, manganese, cobalt, and rare earth elements. Only in the last few years, with improvements in robotics and increasing demand by a world dependent on such materials for electronics components, has the high investment required for seafloor mineral extraction finally achieved a profitable margin. The CEO of the Canadian-based seafloor-mining company DeepGreen enthusiastically describes deep sea minerals as "metals for our future," needed "for a more sustainable planet."⁶⁶ Indeed, remotely operated deep-sea mining machinery is now being tested off the coast of Papua New Guinea.⁶⁷ Meanwhile, a 2018 comparative study of scientific databases found that a single corporation—the world's largest chemical company, Germany-based BASF—now holds 47 percent of the patents registered for gene sequences harvested from marine species.⁶⁸

Some conservationists have turned to science fiction and art to prod public imagination of threats to the future of our oceans. As part of a project called Radical Ocean Futures, marine scientists have presented future ocean scenarios for marine sustainability set in 2070, complete with illustrations of dystopian seascapes by the artist Simon Stålenhag (see Figure 5.4). Project organizer Andrew Merrie, a researcher at the Stockholm University Resiliency Center, explains, "Our scenarios address the contested reality of what counts as a desirable future and for whom. Scenarios can be used to keep track of what sort of future is emerging,

Figure 5.4 "Rime of the Last Fisherman," by Simon Stålenhag. Copyright © Simon Stålenhag; reproduced with permission.

and provide inspiration for how we might steer towards a future that is more equitable and socially and ecologically desirable."[69] But others have abandoned the idea of working toward an improved collective future, instead imagining exclusive libertarian utopias, refuge of a privileged few, afloat beyond national jurisdictions. This is the dream of the Seasteading Institute, founded by a Google engineer in 2008 with financial backing from the billionaire libertarian venture capitalist Peter Thiel.[70]

The reality is that many environmental impacts from climate change remain difficult to predict, especially because, as oceans change, humans modify their behavior. Current environmental changes are particularly evident in the Arctic, where temperatures are increasing at a more rapid rate than anywhere else on the globe. David Titly, a retired naval admiral, testified in a 2018 US House of Representatives hearing on Arctic issues, "We have never been in a position in the modern world where access to an entire ocean opened up within a matter of decades."[71] These changes have geopolitical ramifications because maritime powers vie with one another for control of new shipping passages and seafloor resources. China has already announced plans to send ships through the Northwest

Passage, soon to be ice-free. "There will be ships with Chinese flags sailing through this route in the future," a government spokesman proudly announced in the *China Daily*. "Once this route is commonly used, it will directly change global maritime transportation and have a profound influence on international trade, the world economy, capital flows and resources exploitation."[72] The Northwest Passage promises to save 9000 kilometers (4,860 nautical miles) compared to the traditional route between Europe and Asia via the Panama Canal.[73] Meanwhile, other maritime powers are drawn north by the promise of untapped oil and mineral riches on the seafloor. Russia has launched a floating nuclear reactor, the *Akademik Lomonosov*, and announced plans for further underwater infrastructure development in the Arctic Ocean.[74] Industrial marine infrastructure has been greatly expanded since the 1970s and will continue to grow. Fiber-optic cables crisscross the oceans of the world, offshore windfarms have been built in the North Sea and on the Atlantic Seaboard, and cloud-computing companies have begun to experiment with underwater data centers.[75]

Changing oceans have added new urgency and new directions of study for marine scientists. David Fields, a senior research scientist at Bigelow Laboratory for Ocean Sciences concludes, "The change is happening so fast that it is really a fantastic time to be a scientist, and a horrible time to be a citizen. We are seeing in our lifetime changes that wouldn't normally happen for thousands of years."[76] As this acceleration takes place, much oceanographic fieldwork has shifted from exploration of an unknown frontier to the monitoring, observation, and testing of established models.[77]

There are many signs that marine life too has begun to alter its patterns of behavior in response to the human-caused environmental changes. Blue whales, for example, have changed the pitch of their vocalizations as they strive to overcome the increasing cacophony of ship engine noise.[78] And as on land, many species have responded to changing temperatures by migrating into regions that are more hospitable.[79] On the eastern coast of the United States, the lobster fishery has gradually been moving northward, and the traditional lobstering grounds off the coast of New England are less productive than in the past.[80] In North Carolina's Pamlico Sound, as water temperatures rise bull sharks have begun to appear in unprecedented numbers, using the sound as a nursery for their young.[81] Because the region might now be more hospitable for the species scientists are concerned that juvenile sharks will outcompete other resident species for food, with undesirable results for the entire ecosystem. As a result of changing water temperatures many commercial species are now on the move—posing a problem for the humans whose marine laws have been designed to regulate long-standing commercial uses of specific, bounded marine areas. And rising temperature is but one of myriad ways that human activities have disrupted the habitats of marine creatures. For instance, scientists are now monitoring how underwater electrified cables—used for a broad range of applications—affect the behavior of creatures living on or below the seafloor.[82] And they have begun to investigate how stationary marine infrastructure, like oil rigs and wind farms, create habitats for sessile pelagic creatures (which can in turn affect local primary productivity).[83] Many of the world's exclusive economic zones are estimated to likely receive one to five new, climate-driven transboundary stocks by the end of the century. Some exclusive economic zones in East Asia—already a heavily contested territorial region—could receive up to ten new stocks as commercial species continue to migrate.[84]

The Ocean as Laboratory

In 1957, the Scripps director Roger Revelle famously wrote,

> Human beings are now carrying out a large-scale experiment of a kind that could not have happened in the past nor be reproduced in the future. Within a few centuries we are returning to the atmosphere and oceans the concentrated organic carbon stored in sedimentary rocks over hundreds of millions of years. This experiment, if adequately documented, may yield far-reaching insight into the processes determining weather and climate.[85]

Humanity has since done little to slow the progress of this "experiment" with the oceans. Mounting scientific evidence has confirmed predictions that climate change and overfishing are transforming the world's oceans at an unprecedented rate. In a 2009 interview with the BBC, the marine biologist Carol Turley of the Plymouth Marine Laboratory, speaking about the threat of ocean acidification, warned that this "rate of change in the oceans hasn't been seen since the dinosaurs." "One thing is certain," she added, "it is not a very wise experiment to be making."[86] The oceans are changing as a result of human interference, and the scientific approach to the study of the oceans must adapt as well.

In the Bay of Naples, near the island of Ischia, naturally occurring volcanic seeps release bubbles of carbon dioxide into the water column. This results in localized high concentrations of carbonic acid, making the area extremely inhospitable to organisms with calcareous shells. In marked contrast to nearby ecosystems, seagrass flourishes in the carbon dioxide rich environment. Papua New Guinea has a similar site off its coast. These sites are of particular interest to marine scientists because they are not only exemplary of local environmental conditions but also offer a glimpse of possible future conditions in the oceans as a whole.[87] As climate change refocuses scientific attention to global systems, oceanographic fieldwork is following suit, shifting from a science of predominantly localized observation and analysis to one of global systems prediction.

A good example of this shift in methodology is Geotraces, an international collaborative study of marine biogeochemical cycles of trace elements and their isotopes. Formally launched in 2010, Geotraces has the participation of over thirty countries and is now one of the most important funding sources for chemical oceanographic research. As laid out in the program's science plan, its primary aim is to "identify processes and quantity fluxes that control the distributions of key elements and isotopes in the ocean, and to establish the sensitivity of these distributions to changing environmental conditions." This information, the report claims, offers "unique insights into the future consequences of global change."[88] As of 2019, the ongoing program has provided support for over one hundred research cruises.[89]

As a result of this reorientation to future global-scale changes in the marine environment, some scientists have shifted their attention to the study of marine ecosystem manipulation—experimentation on a grand scale. Algal plankton, minuscule single-celled organisms found in large concentrations in the oceans, account for approximately 50 percent of the photosynthesis on earth. As plankton photosynthesize, as does any terrestrial plant, they remove carbon dioxide from the atmosphere and release oxygen. It is estimated that 99.9 percent of the carbon dioxide that has been incorporated by biological organisms over geologic time is buried in marine sediments. When plankton die off, the remaining organic material—"marine snow"—sinks in the water column, trapping carbon dioxide in the depths of the ocean. This store of carbon is what climate scientists refer to as a "climate sink," and some have proposed that plankton blooms should be artificially induced to mitigate climate change.[90]

Since 1993 at least twelve ocean experiments to artificially trigger plankton blooms have been carried out.[91] Because iron is an important nutrient for algae, marine scientists have hypothesized that simply adding iron dust to the marine environment could trigger large-scale plankton blooms. As the marine biochemist John Martin, former director of the Moss Landing Marine Laboratory, famously quipped, "Give me half a tanker of iron, and I'll give you an ice age."[92] But as another marine geochemist later acknowledged, "[In the past] we were trying to answer the question, 'how does the world work?'—not 'how do we make the world work for us?'"[93] Since the mid-2000s the scientific consensus has shifted to regard iron seeding as too risky and unpredictable to offer a viable solution to climate change.[94] Dreams of radically reshaping the world's oceans through geoengineering seem to have abated, at least for now. Nevertheless, as oceans become more accessible to scientists and as pressure to exploit marine resources to an ever-greater extent builds, many more people, nation-states, and institutions will gain the ability to dramatically shape the oceans' future.

Sensor buoys, remotely operated vehicles, unmanned surface vehicles, and autonomous underwater vehicles have been used to gather data and transfer it nearly instantaneously to laboratories on land via satellite. The Canadian ocean observatory system, the Victoria Experimental Network Under the Sea (VENUS), a fiber-optic network allowing permanent in situ monitoring of oceanic conditions off the coast of British Columbia has been operating since 2006. This was followed by the launch of the North-East Pacific Time-series Undersea Networked Experiments (NEPTUNE) in 2009. In the United States, the Washington regional node of the US

Oceans Observatory Initiative, led by the University of Washington oceanographer John Delaney, finished laying its first submarine cable in 2011. These technological innovations have changed the way geologic, physical, and chemical marine data are gathered at sea. Fiber-optic observatories are now in development off the coasts of the United States, Brazil, Japan, the Azores, and China. Globally, surface layers of the oceans are monitored by a fleet of nearly four thousand autonomous Argo floats, which relay data back to land-based centers via satellites. Deep Argo floats are now being tested to use at depths of up to six thousand meters. The ocean, crisscrossed by sensors and fiber-optic cables, has been reconceived as a "natural laboratory" or "laboratory system."[95]

Scientists have used dedicated ocean-observing satellites since the deployment of Seasat A in 1978.[96] Robots have also successfully been deployed to study regions deemed too dangerous for research vessels. (In 2001 an American research vessel, *Maurice Ewing*, was pursued by pirates off the coast of Somalia.[97]) In other words, although the scale of the ocean is vast, so is the scale of scientific monitoring, and in many ways the ocean increasingly resembles a laboratory.[98] The Argo floats program, established in 2000, achieved its goal of deploying 3000 floats in 2007. By 2019 there were over 3900 floats deployed.[99] Research vessels too have been revolutionized by modern computing and communications technology. The *Okeanus Explorer,* commissioned in 2008, the youngest research vessel in the National Oceanic and Atmospheric Administration fleet, can transmit data (including images and high-definition video) via high-speed internet to scientists and anyone with an internet connection anywhere in the world.

Technological advances have extended the field of scientific observation far beyond the bounds attained by earlier research expeditions. John Delaney observed in 2010, "There will be massive amounts of data flowing ashore, all available to anyone who has any interest in using it. This is going to be much more powerful than having a single ship in a single location."[100] In 2007, Delaney described the potential of remote observatories in even grander terms, "This is a NASA-scale mission to basically enter Inner Space, and be there perpetually. What we're doing is bringing the ocean to the world."[101] Research vessels continue to serve as scientific instruments by deploying remote sensors, yet the ships themselves—supply vessels to a system of sensors embedded in the ocean—are no longer primary collectors or producers of scientific knowledge. In the new ocean-laboratory, oceanographers can more easily present their work as "experiments."[102] In Delaney's words, we are witnessing the "dawning of a new age of how humans can explore the oceans."[103]

But technological innovations are changing not only science at sea; they also have the potential to change marine governance. For this reason, the marine biologist J. Emmett Duffy describes the emerging ocean-scape as "ocean 2.0." Among the most important of twenty-first-century developments, he observes, are the crowdsourcing activities and international scientific collaboration made possible by social media. The nonprofit organization SkyTruth posts near-real-time ship-position information gathered by using satellite tracking; the aim of their Global Fishing Watch program is to use crowdsourcing to identify, and apprehend, illegal fishing vessels anywhere in the world.[104] In 2015, satellite tracking galvanized public opposition to, and halted, seismic surveying operations that threatened coral reefs off the coast of Belize.[105] As a result, in June 2018, the Belize Barrier Reef was officially removed from the UN list of endangered World Heritage sites—a rare victory for marine conservationists.[106] In a sense, ordinary citizens can now be a part of an international ocean police network, albeit one quite different from that envisioned by Carroll Livingston Riker in 1915.

Not all the innovations in remote sensing are intended to monitor human activities. We live in what has been called "the golden age" of animal tracking. Until the mid-twentieth century, the best records available about the location of whales were from areas where whales were hunted. In efforts to gather data from living whales, naturalists adopted highly invasive techniques. The cardiologist Paul Dudley White, for example, tried in 1953 to measure the heartbeat of a California gray whale by inserting subcutaneous electrodes connected to a floating sea sled carrying an electrocardiogram. The whale broke free, and White was forced to admit the experiment a failure.[107] Today, cetacean researchers have available sophisticated tracking devices that not only provide real-time positioning via satellite but also, thanks to magnetometers and accelerometers, record precise movements, all without harming or hindering the animal.[108] These tracking data have been used to build computer simulations that calculate probable animal movement to reduce risk of whale-ship collisions in high-traffic areas.[109] Modern marine scientists have access to data on a scale previously unimaginable. Yet the panoptic reframing of ocean spaces once again suggests a degree of control over marine spaces that, if we heed the lessons of the past, may turn out to be illusory.

Conclusion: "Dying Seas" of the Anthropocene

Declarations that the ocean is dying have become commonplace. We read headlines almost daily telling us that the oceans are choked with plastic, overfished, and rapidly acidifying. Yet even in "dying," we are told, the ocean threatens human existence as sea levels rise, sea surface temperatures increase, and commercial fish stocks disappear. The ocean has thus become emblematic both of a natural world victimized by humanity and of nature's possible vengeance. In a 2014 video by the nonprofit organization Conservation International, the growling baritone of the actor Harrison Ford speaks for the ocean: "I give. They take. But I can always take back."[110] The message is powerful because it conjures images of both the primordial sea as crucible of life and the biblical flood—destruction of life as punishment for human sin. Yet a vengeful ocean is but one of several historical depictions of the sea, some of which have gained prominence at particular moments while others have faded away. In the 1960s and 1970s many scientists, engineers, and policy makers approached the ocean as a vast but resistant reservoir of untapped natural resources. The hostility of the ocean was understood in the context of national calls for increasing exploitation. US Rear Admiral William C. Hushing, for example, in 1967 described the ocean as "hostile in almost every way you can think." In Hushing's view, the task set for "Man" was "to train himself for the hostility" and eventually "find ways to convert the hostility to friendliness."[111]

Today, the ocean is increasingly cast as fragile, even as dying. And while the ocean voiced by Harrison Ford remains threatening, the message is that humans are responsible for that threat. We, not the ocean, have taken too much. Once we recognize the increasing dominance of a conception of the ocean as fragile and dying, we are prompted to ask how this shapes conservation efforts and whether it has a net positive or negative influence on marine environmental protection. In the fall of 2016, for example, *Outside Magazine* published an obituary for the Great Barrier Reef.[112] The article quickly went viral, but coral reef scientists condemned the story as irresponsible.[113] The Great Barrier Reef, they pointed out, although under severe threat, was not yet dead. To declare it lifeless was to give up hope.[114] Environmental pessimism comes at a cost. When pseudoscientific claims gain traction, it is often because they appeal to emotions and long-standing narratives already associated with particular environmental spaces.

Dying-seas narratives and imagery may actually hamper communication between scientists and the public.[115] As an example, Jay Cullen, a researcher at the University of Victoria, leads a project to monitor Fukushima radiation in the eastern Pacific. When Cullen's lab reported that trace radiation was present off the coast of British Columbia but did not represent a significant health hazard, the response was vociferously angry, including death threats aimed at Cullen.[116] In the case of the Fukushima radiation reports, one public's response was to reject scientific claims that did not support the narrative of threatening "dying seas." To quote the *Globe and Mail:* "Dr. Cullen said he frequently hears from people that his science simply can't be right because the Pacific Ocean is dying. It is adrift with tsunami debris and plastic waste and its stocks have been overfished, but it has not been killed by nuclear radiation."[117]

Although hampering science communication, the dying-seas narrative may also contribute to misguided efforts at environmental restoration. In 2012 a native community on Haida Gwaii paid $2.5 million to an American entrepreneur to carry out an iron-seeding experiment off the coast of British Columbia. The goal was to dump iron dust into the sea to artificially trigger a plankton bloom and restore the local salmon population while also sequestering carbon dioxide. As mentioned earlier, oceanographers pioneered iron-seeding experiments but came to deem the method as too risky for practical use. The Haida Gwaii iron-seeding project was therefore condemned by the international scientific community as having violated two international agreements to place checks on unregulated geoengineering. Yet a lay public that was sold on *saving* a "dying sea" triggered what many in the scientific community saw to be dangerous "rogue science."[118] Nor is the 2012 iron-seeding event the only scientifically questionable technological solution marketed as a solution for marine ecological crises. A far more ambitious engineering project to skim microplastics from the North Pacific sea surface is now being tested. The Ocean Cleanup project was founded by a teenage Dutch inventor who, after delivering a viral TEDx speech and raising $2.2 million in crowdsourced funding, dropped out of university to develop his project. Despite concerns voiced by oceanographers that the device will not only be ineffective but will harm pelagic marine creatures, the installation was deployed in late 2018.[119] On a much smaller scale, millions of dollars have been invested in engineering projects around the world in the Sisyphean task of trying to hold back rising seas as the Greenland and Antarctic ice sheets melt.[120] It may be that future oceanographers, unlike their predecessors, will be less focused on encouragement of widespread collaborative observation and experimentation at sea and more concerned with oversight and restriction of interfering scientific and engineering practices.

Unsurprisingly, the projection of sentience onto the natural world fails to move climate change skeptics. Appeals to safeguard individual charismatic species, like the polar bear, risk critique as devaluing human existence in favor of other forms of life.[121] Descriptions of the earth as a victim of human agency are dismissed by political opponents as scientific hubris.[122] Even publics potentially receptive to conservation science risk being demoralized by imaginative invocation of a vast, "dying" nonhuman entity. The author of a 2014 editorial in *Smithsonian Magazine* notes, "We've gone from thinking the ocean was too big to hurt, to thinking that the ocean is too big and too sick to help."[123] This cognitive-emotional orientation has been unintentionally fostered by scientists intent on educating a lay public on the importance of global systems thinking. Yet the popularization of this approach to nature has its pitfalls. Conceptualizing the oceans as a cohesive nonhuman entity oversimplifies accounts of environmental degradation and limits understanding of local variability.

In 2013, Microsoft cofounder Paul Allen announced a contest called Ocean Challenge. The contest awarded "$10,000 to the most promising new science-based concept for mitigating environmental and/or societal impacts of ocean acidification." The winners of the contest were Ruth D.

Gates of the University of Hawaii and Madeleine van Oppen of the Australian Institute of Marine Science. Their project to genetically select and cultivate corals that possess natural resistance to ocean acidification received funding. Coral reefs take up less than 1 percent of the earth's surface, yet they are habitats for an estimated one-third of all known marine creatures, including 25 percent of commercial seafood species. They also act as natural breakwaters, dampening the power of storm surges and coastal erosion. An estimated 61 percent of coral reefs are under stress and at risk of disappearing by 2030. Thus, the health of coral reefs is widely used as a metric for global ocean health, the marine equivalent of the canary in the coal mine. Gates, who passed away in October 2018, described herself as "a futurist." "A lot of people want to go back to something. They think, If we just stop doing things, maybe the reef will come back to what it was," she explained. In contrast, her project acknowledged a future "where nature is no longer fully natural."[124] In Gates's understanding, the ocean isn't dead, but its survival hinges on assumption of responsibility for its now-hybrid character. Is there a cost to abandoning the nineteenth-century ideal of wilderness? Perhaps doing so is the price we must pay to retain a semblance of what once was.

Some theorists and scientists advocate greater inclusion of nonhuman actors in debates about ecological crisis. Bruno Latour, for example, argues that "a science of objects and politics of subjects" must be replaced by a "political ecology of collectives consisting of humans and nonhumans."[125] A precedent has been set by the recent allocation of legal rights to rivers in Australia, New Zealand, and India.[126] But although we must not shirk from placing value on nonhuman entities, in the end climate change—and by extension marine environmental degradation—remains a human problem, and we need to foreground human abilities to comprehend and solve it. As Jean-Michel Cousteau, son of Jacques, asserts, "The face of our planet is the ocean. It is the largest ecosystem on our Earth. But the face of climate change is not the whale, the polar bear, the glacier, the rainforest or the desert. The face of climate change is us."[127]

The marine sciences, like all branches of scientific knowledge, are shaped by underlying assumptions about human relationship with the natural world. The tensions I have highlighted point to a crisis in scientific and lay imaginations of an ocean radically changed in the course of the Anthropocene. Scientists increasingly talk about the ocean as a hybrid environment.[128] Gates was surely correct in asserting that scientific solutions for an ocean understood as dying can be reached only by acknowledging that the contemporary ocean cannot be conceived apart from humans. But even this perspective comes with risk. We see the debate playing out in discussions over optimal approaches to fisheries management—between scientists who advocate "no-take," wild marine protected areas and others arguing that proper management protects biodiversity through judicious harvesting.[129] History suggests that every manipulation with nature will have unpredictable effects.[130] It may be hubris to think that we can cure a condition whose root cause is ourselves.

Humans live in symbiosis with the sea. Even those who choose not to eat seafood or do not live near a seacoast might be surprised to learn the extent to which their livelihood, health, and well-being depend on the ocean. For example, 25 percent of all global fish catch ends up as agricultural fertilizer, food for domestic livestock, fish oil supplements, or processed food for commercially farmed fish.[131] Carrageenan extracted from seaweed is used to thicken ice cream and yogurt, and offshore oil wells account for approximately 30 percent of global oil production. Most importantly, marine plankton produce 70 percent of the oxygen in the atmosphere. The famed oceanographer Sylvia Earle frames our dependence on the ocean in existential terms: "nothing else will matter if we fail to protect the ocean. Our fate, and the ocean's, are one."[132]

It remains imperative for those seeking to mobilize support for efforts of ocean conservation to understand underlying conceptions that shape how scientists and lay publics approach the natural world. Climate change is giving rise to novel fears about rising sea levels, slowing thermohaline circulation, and acidifying seas. Some dire predictions made in the past have been borne out sooner than anticipated. Yet prediction is inherently imperfect. If we are not careful, fear may hamper conservation efforts or fuel hubristic interventions.[133] Hope, like fear, has power to shape the world we will inhabit.

Conclusion

We must live in hopes, supposing we die in despair.
Newfoundland proverb

I look at the role of imagination in the history of the marine sciences in this book. Surveying developments from the nineteenth century up to the present, we note the recurrence of certain hopes, as well as fears, expressed by scientists, politicians, and members of the public. The vast ocean-scape, defying in its scale the scientific resources of any single individual or nation, has for scientists long conjured dreams of building extended collaborative networks. This aspiration stimulated the growth and development of a branch of science conducted in the periphery, away from powerful institutional centers. It gave impetus to both national and international marine exploration projects. On the national scale, we can point to Edward Forbes's dredging committee in Britain in the early nineteenth century, and on an international scale, to the efforts of the American scientist Matthew Fontaine Maury to organize an international maritime conference in the 1850s. Such projects were given further momentum by an intellectually curious public, fascinated by the wonders of the sea, and by politicians who viewed marine science as a way to promote national honor and glory on an international stage.

We have seen from the examples of Henri Lacaze-Duthiers and Prince Albert I of Monaco how, in the second half of the nineteenth century, dreams of international collaboration were stunted by growing nationalism and fears of military conflict. World War I weakened incipient internationalist marine science programs in the North Sea, the Mediterranean, and the Atlantic. The less well studied, and seemingly virgin, region of the Pacific Ocean, however, remained a frontier where dreams of international collaboration continued to flourish. Hence, in the first part of the twentieth century we once again find calls for international collaboration, this time in the scientific exploration of the Pacific basin. And once again, not only was this dream fueled by scientists hoping to build collaborative networks; it was also supported by a public eager to establish a new identity—removed from the violence of the Old World—as citizens of a Pacific World.

This internationalist dream diminished with the start of World War II, because oceanography was repurposed to serve military goals, but it was not fully extinguished. We find at the fairground—both in San Francisco in the 1930s and in Seattle in the 1960s—continued evidence of a presentation of science as a means for achieving peace, international harmony, and a better human future. This optimistic vision impelled amateur explorers to seek out ways of participating in marine science and exploration, even while politicians and military strategists planned the increasing militarization of the oceans. Nevertheless, it was precisely dread of this militarization, as well as anxieties about industrial exploitation of the seabed, that stimulated alternative utopian dreams of new forms of governance for oceanic space and a new relationship to a "humanized ocean"—such as Arvid Pardo's and Elisabeth Mann Borgese's.

In the present day, we are once again confronting the specter of environmental collapse and ensuing threats to civilization. Although we can hazard some guesses about how our responses might shape or hinder policy, it is still too soon to know how they will shape humanity's future and the ocean's. Having a better understanding of how decisions, both scientific and political, are inevitably molded by hopeful and fearful imagination will undoubtedly help us face the coming challenges.

New Directions

We are still in the early years of the oceanic turn in the humanities, and there remain many possibilities for fruitful work. As we have seen, the historical figures that are often easiest to follow are the scientists.[1] But as we learned from the example of recreational scuba divers in the twentieth century, more can be done to reveal the roles and contributions of amateur practitioners, nonacademics, and other hidden figures.[2] Furthermore, the narrative I have presented has focused largely on developments in the West, especially in Britain, France, and the United States. Other national and peripheral contexts remain to be explored. We saw, for example, how scientific collaborations arose in the pan-national Pacific World of the first half of the twentieth century. Such transnational perspectives on the history of marine science will prove ever more important as the global environmental crisis sets problems beyond the capability of any single nation-state. Finally, there remain large gaps in our knowledge of the history of the marine sciences. Critically, the history of oceanography in the Soviet Union remains to be written and would fill an important lacuna in the existing scholarship.

In the Anthropocene we face unprecedented pressures on the environment and threats to global biodiversity. But we also live in an era of renewed nationalism and increasing militarization of the oceans. Cold War rivalries are beginning to reheat with additional new states in play. As of 2017, China possesses the largest navy in the world and has aggressively extended territorial claims by building artificial islands and by restricting freedom of navigation in disputed waters.[3] The same technology that has permitted a better understanding of the oceans, robotics and drones, is rapidly being adapted for military purposes.[4] As marine spaces become more accessible to military and industrial interests, international rivalries will increase. We stand at a crossroads. In September 2018 the United Nations began negotiations toward the first-ever international treaty for the protection and management of the high seas. And despite opposition from some countries, these efforts are ongoing. At this moment in history we have something to gain from remembering, and even trying to recover, the optimistic internationalist vision of some who came before us.

Living in Hope

In 2000, President Bill Clinton signed bipartisan legislation to establish a national advisory board on ocean protection and ocean-dependent industries. As he said then, "We know that when we protect our oceans we're protecting our future."[5] Subsequent presidents have followed suit.

President George W. Bush established what was, at the time, the world's largest ocean preserve, in remote islands of the Hawaiian chain, an area of 84 million acres (approximately 131,250 square miles). In 2014 President Barack Obama expanded the Pacific Remote Islands National Monument, and in 2016 he quadrupled the size of the Papahanaumokuakea Marine National Monument established by Bush. Although the Donald Trump administration has carried out an unprecedented rollback of environmental protections, it has not shrunk existing marine protected areas. Yet.

The 2010 Deepwater Horizon disaster underscored to politicians, fishers, and tourism operators the danger of taking marine resources for granted. The Trump administration's efforts to open regions along the Eastern Seaboard to offshore drilling have, thus far, been met by bipartisan opposition. Such united resistance, as the former chair of the South Carolina GOP, Matt Moore, described it, is "a strange, but beautiful, thing to see."[6] The public is gradually coming to understand that efforts to protect the oceans also safeguard the people, and other countries are taking steps to protect fragile marine areas. In 2017, Mexico created a marine reserve off its southwest coast the size of Illinois (58,000 square miles). In 2018, Chile announced the creation of three marine parks in the Pacific, covering a total area of 450,000 square miles. And in 2018 the Canadian Department of Fisheries and Oceans proposed the creation of a marine protected area off the coast of British Columbia (approximately 53,000 square miles). Conservationists argue that at least 30 percent of the world ocean must be protected to safeguard marine biodiversity, support fisheries, and sustain the economic, cultural, and life-supporting benefits of healthy seas.[7] We clearly have a long way to go, but the progress made this century should be read as cause for cautious optimism.

Despite the oceans' importance to the survival and well-being of life on our planet, the marine environment remains an area that the majority of us rarely see; when we do, it is but a partial glimpse from the shoreline or the surface. Scientific understanding of the marine environment has greatly advanced since the first dredging expeditions of the nineteenth century. As instrumentation improves, the oceans become more accessible to scientists.[8] And scientific understanding of marine systems is essential to their effective conservation. The United Nations has declared

2021–2030 a Decade of Ocean Science for Sustainable Development. As announced on the UNESCO website, the aim is to deliver "the ocean we need for the future we want!"[9]

If conservationists are to succeed, they must find effective means to educate the public about this remote and often hidden terrain. Making people care about something they cannot experience firsthand is a challenge; outreach by scientists, museums, aquariums, and various forms of media communication is instrumental to this effort. Already there are worldwide examples of marine conservation success stories, places the marine biologist Sylvia Earle terms "hope spots."[10] The popular 2017 documentary series *Blue Planet Two,* produced by the BBC and narrated by David Attenborough, helped launch a global campaign against marine plastics pollution and prompted a noticeable increase in university marine biology enrollments.[11]

We should recognize from the history of marine science that knowledge about our oceans has been increased through many endeavors. The diversity of voices in the history related here suggests the breadth of publics that are potentially receptive to debates about the future of the oceans. The marine microbiologist Julie Huber responded when asked how the "average" person who cares about marine science could help, "You don't need a PhD. . . . We're all inquisitive people. People get scared off by this elitist ivory tower view of scientist, but there's lots of ways to participate in the process as an individual, as a community."[12]

The humanities have a major role to play. In *The Great Acceleration,* J. R. McNeill and Peter Engelke criticize academic social scientists and humanists for having retreated "from the grimy and greasy realities" of environmental destruction in the Anthropocene "into various never-never lands." This "intellectual flight from reality," they argue, has made it "easier for those in positions of power to avoid facing up to it."[13] The rise of the environmental movement, and the subsequent growth of the subdiscipline of environmental history, has opened new opportunities for interdisciplinary dialogue and collaboration. It falls to marine environmental historians, and historians of marine science, to remind us of what has been, what has been lost, and how past conservation measures fared, lending historical insight to contemporary policy deliberations. As William Cronon phrased it, "If our histories are to help change the world, they must reach beyond the walls of the academy to affect the views of people who do more than just study the past."[14]

Putting this advice into action is challenging but imperative. The historian Nancy Langston writes in her introduction to a special issue of the journal *Environmental History* devoted to the marine environment,

> It's not always easy for historians to participate in scholarly and policy debates over marine management. Scientists, policymakers, and historians speak different languages and control different financial resources, which makes collaboration challenging. When a scientist wins a grant and invites a historian to participate in a research project, what the scientist needs from that historian often differs from what the historian wants to contribute. Historians excel at problematizing scientific approaches to knowledge building, but "Well . . . it's complicated" is rarely a useful answer when a policymaker asks a question. Nevertheless, it is critical for historians to participate—and participate usefully—in interdisciplinary marine research. Without a sense of history, how can we hope to understand, much less restore, marine ecosystems?[15]

Interdisciplinary collaboration expands the reach of science communication. This alone is a worthy goal. In the nineteenth century, marine naturalists produced monograph narratives of scientific voyages for public consumption, and readers were enraptured by the mysteries of the abyss and by the science and technology that made such discoveries possible. We may be unintentionally limiting ourselves if we view interdisciplinary collaboration only as a means for solving pressing environmental problems. Interdisciplinary collaboration between scientists and historians can be part of a multipronged effort to reengage the public in marine science and ocean conservation.

In taking on this mission, we can benefit from an appreciation of the role imagination has always played in marine science. Not all past imaginations about the marine environment have been optimistic; but there are lessons to be learned from the periods when fear and nationalism overcame hope and trust and scientific collaboration faltered. Some of our collective anxieties have persisted, and some are well warranted. Fears can serve as useful goads, rousing humans to change course.

Yet we may also take encouragement from the persistent resurgence of utopian aspirations throughout the history of marine science. Periods of geopolitical conflict have failed to extinguish the dream of international political and scientific cooperation at sea. In the twenty-first century the hope for collaborative global marine conservation is finding fulfillment in new institutionalized forms. The ocean can still serve as a canvas on which the best impulses of our nature may be projected and recognized. The lesson is not that dreams are no more than fantasy. Rather, we learn from the history of marine science that imagination can fuel projects of grand scope, ambition, and achievement. Only by first imagining a desirable future will we be able to find a path to protecting our Anthropocene ocean and life, which depends on it.

Notes

Acknowledgments

Index

Notes

Introduction

Epigraphs: Paul Valéry, *Mer, Marines, Marins* (Paris: Firmin-Didot, 1930), 3; Henry W. Menard, *Anatomy of an Expedition* (New York: McGraw-Hill, 1969), 3.

1. Iwan Rhys Morus, "Future Perfect: Social Progress, High-Speed Transport and Electricity Everywhere—How the Victorians Invented the Future," *Aeon* (10 December 2014), https://aeon.co/essays/how-the-victorians-invented-the-future-for-us.
2. Peter Fritzsche, *Stranded in the Present: Modern Time and the Melancholy of History* (Cambridge, MA: Harvard University Press, 2004), 5.
3. Thomas Malthus's *An Essay on the Principle of Population* was first published in 1798.
4. Denise Phillips, *Acolytes of Nature: Defining Natural Science in Germany, 1770–1799* (Chicago: University of Chicago Press, 2012), 9.
5. Any account of the founding of the history of oceanography should acknowledge the prominent role of archivists and librarians who became leading subject specialists; much scholarship would not exist without their efforts. I give particular credit to Jacqueline Carpine-Lancre, archivist at the Oceanographic Museum in Monaco; Deborah Day at the Scripps Institution of Oceanography; Christiane Groeben, archivist at the Naples, Italy, Zoological Station; Ruth Davis, archivist at the Marine Biological Laboratory in Woods Hole, Massachusetts; and Anita McConnell, senior curator at the London Science Museum.
6. Christiane Groeben, introduction to *Places, People, Tools: Oceanography in the Mediterranean and Beyond*, ed. Christiane Groeben (Naples, Italy: Giannini Editore, 2013), 11.
7. Eric Mills, "What Is the History of Oceanography?," *History of Oceanography*, no. 2 (July 1990): 2–3.

8. See, for example, Georg Wüst, "The Major Deep-Sea Expeditions and Research Vessels 1873–1960," in *Progress in Oceanography*, ed. Mary Sears (Oxford: Pergamon, 1964), 2:1–52.
9. Helen Rozwadowski, "Scientists Writing and Knowing the Ocean," in *The Sea and Nineteenth-Century Anglophone Literary Culture*, ed. Steve Mentz and Martha Elena Rojas (London: Routledge, 2016), 28–46.
10. Overlapping with maritime and naval history, histories of oceanography centered on ships and expeditions often focus on particular national contexts. Margaret Deacon, in her now-classic 1971 account of the early development of marine science, critiqued this tendency, arguing that to cite the HMS *Challenger* expedition as the origin for marine science was "to misjudge its significance," neglect "the work of earlier scientists," and underestimate "the strength of the parallel movement in other countries." Margaret Deacon, *Scientists and the Sea, 1650–1900: A Study of Marine Science* (London: Academic Press, 1971), 368–369.
11. The need to incorporate ships in histories of oceanography was one of the conclusions of the Maury Workshop on the History of Oceanography (Maury III), held in June 2001 in Monterey, California. See Keith Benson, Helen Rozwadowski, and David K. van Keuren, introduction to *The Machine in Neptune's Garden: Historical Perspectives on Technology and the Marine Environment*, ed. Helen Rozwadowski and David K. van Keuren (Sagamore Beach, MA: Science History Publications, 2004), xxvi, note 13.
12. On these grounds alone, oceanography falls within a categorical "way of thinking" often neglected by historians of science, particularly in studies dealing with twentieth-century developments. As Vanessa Heggie writes, "The history of twentieth-century science is routinely written about (and taught) without much consideration of nonlaboratory sciences, and the role of extraordinary encounters between human bodies and the earthly environment is rarely discussed." Vanessa Heggie, "Why Isn't Exploration a Science?," *Isis* 105, no. 2 (June 2014): 319.
13. See, for example, the edited volume "Science in the Field," ed. Henrika Kuklick and Robert E. Kohler, *Osiris* 11, no.1 (1996); *Knowing Global Environments: New Historical Perspectives on the Field Sciences*, ed. Jeremy Vetter (New Brunswick, NJ: Rutgers University Press, 2011); Jeremy Vetter, *Field Life: Science during the Railroad Era* (Pittsburg: University of Pittsburg Press, 2016).
14. This study does not closely examine developments in Germany and Scandinavia. For an analytic approach similar to mine but focused on Germany, see Ole Sparenberg, "The Oceans: A Utopian Resource in the 20th Century," *Deutsches Schiffahrtsarchiv* 30 (2007): 407–420. For marine science in Norway, see Vera Schwach, "The Sea around Norway: Science, Resource Management, and Environmental Concerns, 1860–1970," *Environmental History* 18, no. 1 (January 2013): 101–110. For an account of Danish oceanography, see Bo Poulsen, *Global Marine Science and Carlsberg: The Golden Connections of Johannes Schmidt (1877–1933)* (Leiden, Netherlands: Koninklijke Brill, 2016). For the history of physical oceanography and fisheries research in Canada, see Eric Mills, *The Fluid Envelope of Our Planet: How the Study of the Ocean Currents Became a Science* (Toronto: University of Toronto

Press, 2009); and Jennifer Hubbard, *A Science on the Scales: The Rise of Canadian Atlantic Fisheries Biology, 1898–1939* (Toronto: University of Toronto Press, 2006).

15. See the essays by Helen Rozwadowski, Michael Reidy, Jacob Hamblin, Jennifer Hubbard, and Naomi Oreskes in "Focus: Knowing the Ocean: A Role for the History of Science," *Isis* 105, no. 2 (June 2014): 335–391. See also the collection of essays by Michael Chiarappa, Matthew McKenzie, Brian Payne, Victoria Penziner Hightower, Joseph E. Taylor, III, Loren McClenachan, Jennifer Hubbard, Vera Schwach, Christine Keiner, Paul Hom, Marta Coll, Allison MacDiarmid, Henn Ojaveer, and Bo Poulson in "Marine Forum," *Environmental History* 18, no. 1 (January 2013): 3–126.

16. I refer, in particular, to the work of Eric Mills, Helen Rozwadowski, Jacob Darwin Hamblin, Jennifer Hubbard, Michael Reidy, Keith Benson, Gary Weir, and Naomi Oreskes.

17. Elizabeth Deloughrey, "Submarine Futures of the Anthropocene," *Comparative Literature* 69, no. 1 (2017): 32. Helen Rozwadowski, "Oceans: Fusing the History of Science and Technology with Environmental History," *A Companion to American Environmental History*, ed. Douglas Cazaux Sackman (Oxford: Blackwell, 2010), 456.

18. See also Helen Rozwadowski, "The Promise of Ocean History for Environmental History," *Journal of American History*, June 2013, 136–139.

19. See, for example, the work of Jennifer Hubbard, Jeffrey Bolster, and Matthew McKenzie, Callum Roberts, and Carmel Finley; also "Marine Forum," *Environmental History* 18, no. 1 (January 2013). There is overlap here too with the work of historical ecologists; see Kathleen Schwerdtner Mánez, Paul Holm, Louise Blight, Marta Coll, Alison MacDiarmid, Henn Ojaveer, Bo Poulsen, and Malcolm Tull, "The Future of the Oceans Past: Towards a Global Marine Historical Research Initiative," *PLOS One* 9, no. 7 (2014), https://doi.org/10.1371/journal.pone.0101466.

20. John R. Gillis and Franziska Torma, introduction to *Fluid Frontiers: New Currents in Marine Environmental History*, ed. John R. Gillis and Franziska Torma (Cambridge: White Horse Press, 2015), 1–12.

21. W. Jeffrey Bolster, "Putting the Ocean in Atlantic History: Maritime Communities and Marine Ecology in the Northwest Atlantic, 1500–1800," *American Historical Review* 113, no. 1 (February 2008): 20. However, this understudied gap in environmental history is rapidly closing. See Kelly P. Bushnell, "Oceans," in *Victorian Literature and Culture* 46, nos. 3–4 (Fall/Winter 2018): 788–791; Steven Mentz, "Toward a Blue Cultural Studies: The Sea, Maritime Culture, and Early Modern English Literature," *Literature Compass* 6, no. 5 (2009): 997–1013.

22. For an interesting example of this approach, see Ryan Tucker Jones, "Running into Whales: The History of the North Pacific from Below the Waves," *American Historical Review* 118, no. 2 (1 April 2013): 349–377.

23. Helen Rozwadowski, Robert Marc Friedman, Michael Reidy, and James Roger Fleming are among the scholars who have contributed to this new focus.

24. Robert Kohler, "History of Field Science: Trends and Prospects," in Vetter, *Knowing Global Environments*, 216. For an example of this approach, see Helen Tilley, *Africa*

as a Living Laboratory: Empire, Development, and the Problem of Scientific Knowledge, 1870–1950* (Chicago: University of Chicago Press, 2011), 12.
25. Robert Kohler, *Landscapes and Labscapes: Exploring the Lab-Field Border in Biology* (Chicago: University of Chicago Press, 2002).
26. See, for example, Frank Winter, *Prelude to the Space Age: The Rocket Societies* (Washington, DC: Smithsonian Institution Press, 1983); and Howard E. McCurdy, *Space and the American Imagination* (Washington, DC: Smithsonian Institution Press, 1997).
27. W. Patrick McCray, *The Visioneers: How a Group of Elite Scientists Pursued Space Colonies, Nanotechnologies, and a Limitless Future* (Princeton, NJ: Princeton University Press, 2013).
28. Sheila Jasanoff, "Future Imperfect: Science, Technology, and the Imaginations of Modernity," in *Dreamscapes of Modernity: Sociotechnical Imaginaries and the Fabrication of Power*, ed. Sheila Jasanoff and Sang-Hyun Kim (Chicago: University of Chicago Press, 2015), 4.
29. Early marine naturalists were, almost without exception, male. Marine scientists, immersed in the cultural traditions of the male-dominated maritime world, have long upheld social barriers against the inclusion of women. This practice persisted into the twentieth century. More work remains to be done by historians to recover the voices of the pioneering women who persevered in the face of discrimination. Two memoirs written by female oceanographers are Helen Raitt, *Exploring the Deep Pacific* (New York: Norton, 1956); and Kathleen Crane, *Sea Legs: Tales of a Woman Oceanographer* (Boulder, CO: Westview, 2003).
30. As Jacob Hamblin writes, in the first decades after World War II, "support for [oceanographic] research was based on its usefulness for making war on other nations. At the same time oceanography retained an identity that tied it closely to international cooperation." Jacob Hamblin, *Oceanographers and the Cold War: Disciples of Marine Science* (Seattle: University of Washington Press, 2005), xviii. See also Gary Weir, *An Ocean in Common: American Naval Officers, Scientists, and the Ocean Environment* (College Station: Texas A&M University Press, 2001).
31. Hamblin, *Oceanographers and the Cold War*, xix.
32. Elizabeth Noble Shor, *Scripps Institution of Oceanography: Probing the Oceans, 1936 to 1976* (San Diego, CA: Tofua Press, 1978), 3.

1. Discovering Wonder in the Deep

1. Helen Rozwadowski, *Fathoming the Ocean: The Discovery and Exploration of the Deep Sea* (Cambridge, MA: Harvard University Press, 2005), 6.
2. Katharine Anderson, *Predicting the Weather: Victorians and the Science of Meteorology* (Chicago: University of Chicago Press, 2005), 2.
3. Richard Sorrenson, "The Ship as a Scientific Instrument in the Eighteenth Century," *Osiris* 11 (1996): 223.
4. Robert Boyle, "Relations about the Bottom of the Sea," *Tracts about the Cosmicall Qualities of Things* (1672), 1. See the discussion of Boyle's work in Margaret Deacon, *Scientists and the Sea, 1650–1900: A Study of Marine Science* (Brookfield, VT:

Ashgate, 1997), 117–129. See also Steven Shapin, *A Social History of Truth: Civility and Science in Seventeenth-Century England* (Chicago: University of Chicago Press, 1994).
5. Boyle, "Relations about the Bottom of the Sea," 2.
6. Anita McConnell, *No Sea Too Deep: The History of Oceanographic Instruments* (Bristol, UK: Adam Hilger, 1982), 6–7.
7. Edmond Halley, "An Historical Account of the Trade Winds, and Monsoons, Observable in the Seas between and near the Tropicks, with an Attempt to Assign the Physical Cause of the Said Winds," *Philosophical Transactions of the Royal Society* 16 (1 January 1687): 162.
8. Alan H. Cook, *Edmond Halley: Charting the Heavens and the Seas* (Oxford: Clarendon Press, 1998), 264.
9. James Delbourgo, "Underwater-Works: Voyages and Visions of the Submarine," *Endeavour* 31, no. 3 (2007): 116.
10. McConnell, *No Sea Too Deep*, 12.
11. John Stoye, *Marsigli's Europe 1680–1730: The Life and Times of Luigi Ferdinando Marsigli. Soldier and Virtuoso* (New Haven, CT: Yale University Press, 2004), 27. For more on Marsigli, see F. C. W. Olson and Mary Ann Olson, "Luigi Ferdinando Marsigli, the Lost Father of Oceanography," *Quarterly Journal of the Florida Academy of Sciences* 21, no. 3 (September 1958), 227–234; and Anita McConnell, introduction to *Natural History of the Sea by Luigi Ferdinando Marsigli* (Bologna, Italy: Museo di Fisica dell Universita di Bologna), 6–28.
12. Stoye, *Marsigli's Europe 1680–1730*, 124.
13. Louis Ferdinand, Comte de Marsilli [sic], *Histoire Physique de la Mer* (Amsterdam, 1725), 2. All translations are mine unless otherwise noted. There are many different spellings of Marisigli's name in the archival record (Marsigli, Marsilli, Marsili, or Marsilly).
14. As Anita McConnell points out, however, Marsigli sometimes omitted from his published work any mention of the contributions of other scientists with whom he corresponded and collaborated. McConnell, *No Sea Too Deep*, 11.
15. McConnell, 11.
16. See Joyce E. Chaplin, "Knowing the Ocean: Benjamin Franklin and the Circulation of Atlantic Knowledge," in *Science and Empire in the Atlantic World*, ed. James Delbourgo and Nicholas Dew (New York: Routledge, 2008), 73–96.
17. There are undoubtedly exceptions, but they are rare. The British polar explorer Luke Fox, for example, conducted soundings at sea in 1631. Margaret Deacon, "Founders of Marine Science in Britain: The Work of the Early Fellows of the Royal Society," *Notes and Records of the Royal Society of London* 20, no. 1 (June 1965): 30.
18. Johann Reinhold Forster, *Observations Made during a Voyage Round the World*, ed. Nicholas Thomas, Harriet Guest, and Michael Dettelbach (Honolulu: University of Hawaii Press, 1996), 78. Forster did include in his final report a series of seawater temperature measurements and the positions at which they were taken, although it is unclear whether he made these measurements himself. Margaret Deacon suggests that the expedition's astronomers, William Wales and William

Bayly, made the temperature measurements. Deacon, *Scientists and the Sea*, 186–188.

19. Sorrenson, "Ship as Scientific Instrument," 227. See also J. C. Beaglehole, *The Life of Captain James Cook* (Palo Alto, CA: Stanford University Press, 1974), 293. As Beaglehole explains, "[Cook's ship] was not chosen as a passenger ship or a floating laboratory or as an artist's studio, but precisely because she was what she was—a soundly-built collier, with adequate room for her crew and her stores."

20. Charles Darwin, *The Beagle Letters*, ed. Frederick Burkhardt (Cambridge: Cambridge University Press, 2008), 383.

21. See Alistair Sponsel, "An Amphibious Being: How Maritime Surveying Reshaped Darwin's Approach to Natural History," *Isis* 107, no. 2 (2016), 254–281.

22. This is not to diminish the difficulties of land-based collecting in the late eighteenth century. Forster likened botanizing in New Zealand, while threatened by attack from hostile natives, to "pulling burning embers out of a fire." Quoted in Glyn Williams, *Naturalists at Sea: Scientific Travelers from Dampier to Darwin* (New Haven, CT: Yale University Press, 2013), 106. For the difference between hydrography and oceanography, see Rozwadowski, *Fathoming the Ocean*, 34.

23. Thomas Henry Huxley, "Science at Sea," *Westminster Review* 5 (January 1854): 58.

24. Huxley, 58. Huxley drew from his experience as assistant surgeon aboard HMS *Rattlesnake* (1846–1850).

25. Huxley, 60.

26. Charles Wyville Thomson, *The Depths of the Sea [...]* (London: Macmillan, 1874), 49.

27. Lincoln Paine, *The Sea and Civilization: A Maritime History of the World* (New York: Vintage, 2013), 508.

28. "Traveling by Steam," *Baltimore Sun*, 14 July 1838, 4.

29. Between 1815 and 1930, 56 million Europeans emigrated. Paine, *Sea and Civilization*, 528.

30. Paine, 528.

31. "Summer Cruise to the Mediterranean," *Liverpool Mercury*, 29 July 1868, 6.

32. Rozwadowski, *Fathoming the Ocean*, 119.

33. Bernard S. Finn, *Submarine Telegraphy: The Grand Victorian Technology* (London: Science Museum, 1973).

34. Rudyard Kipling, "The Deep-Sea Cables," *Rudyard Kipling: Complete Verse* (New York: Anchor Press, 1989), 173.

35. As Margaret Deacon rightly points out, however, "The facile view that the introduction of the submarine telegraph advanced the study of marine science does not take into account the fact that many deep sea soundings were made as much as a decade before surveys for it began." Deacon, *Scientists and the Sea*, 298.

36. "London, Friday, August 6, 1858," *Times* (London), 6 August 1858, 8.

37. Charles Darwin, *On the Origin of Species by Means of Natural Selection, or the Preservation of Favoured Races in the Struggle for Life*, 6th ed. (London: Odhams, 1872), 344.

38. Susan Schlee, *The Edge of an Unfamiliar World: A History of Oceanography* (New York: E. P. Dutton, 1973), 93.

39. As both Margaret Deacon and Eric Mills have shown, interest in marine biology long superseded interest in the physical conditions of the sea.
40. W. H. Harvey, *The Sea-Side Book: Being an Introduction to the Natural History of the British Coasts* (London: John Van Voorst, Paternoster Row, 1849), 116.
41. The deeper the dredge descended, the stronger and heavier the attached rope needed to be. And because dredging scooped up mud and stone along with marine life, the heavier the equipment, the more difficult it was to recover samples.
42. In his classic 1923 history of marine science, the oceanographer William Herdman devoted the first chapter to the life and work of the Manxman Edward Forbes. Although a foundational work in the history of oceanography, Herdman's account is decidedly Anglocentric. William A. Herdman, *Founders of Oceanography and Their Work: An Introduction to the Science of the Sea* (London: Edward Arnold, 1923). Herdman (1858–1924) worked in the *Challenger* office after graduating from the University of Edinburgh in 1879.
43. Edward Forbes, *The Natural History of the European Seas* (London: John Van Voorst, Paternoster Row, 1859), 10.
44. For instance, the French scientist Georges Aimé (1813–1846) retrieved animals from a depth of just over a mile (1800 meters) while working off the coast of Algeria in the early 1840s. Deacon, *Scientists and the Sea*, 282.
45. Charles Darwin to J. S. Disnurr, 13 June 1851, *The Darwin Project*, https://www.darwinproject.ac.uk/letter/DCP-LETT-1436.xml.
46. Philip Rehbock, "The Early Dredgers: 'Naturalizing' in British Seas, 1830–1850," *Journal of the History of Biology* 12, no. 2 (1979): 328.
47. Rehbock, 332.
48. Rehbock, 349.
49. Rehbock, 349.
50. The publication of Darwin's *On the Origin of Species* in 1859 also furthered interest in deep-sea biota. More evidence refuting the azoic theory was gathered during the *Lightning* expedition in 1868. Jennifer Hubbard, *A Science on the Scales: The Rise of Canadian Atlantic Fisheries Biology, 1898–1939* (Toronto: University of Toronto Press, 2006), 20.
51. Jason Smith, *To Master the Boundless Sea: The U.S. Navy, the Marine Environment, and the Cartography of Empire* (Chapel Hill: University of North Carolina Press, 2018), 133.
52. Schlee, *Edge of an Unfamiliar World*, 23–24.
53. Antony Adler, "From the Pacific to the Patent Office: The U.S. Exploring Expedition and the Origins of America's First National Museum," *Journal of the History of Collections* 23, no. 1 (2011): 49–74.
54. See Antony Adler, "The Capture and Curation of the Cannibal 'Vendovi': Reality and Representation of a Pacific Frontier," *Journal of Pacific History* 49, no. 3 (2014): 255–282.
55. Maury accepted a position on the Exploring Expedition but resigned during disputes over the command mission. See Chester G. Hearn, *Tracks in the Sea: Matthew Fontaine Maury and the Mapping of the Oceans* (New York: International Marine, McGraw Hill, 2002), 73–77. *The Physical Geography of the Sea* (New York: Harper &

Brothers, 1855) was widely read, and not only by scientists. As Peter H. Kylstra and Arend Meerburg have shown, many passages in *Twenty Thousand Leagues under the Sea* were inspired by Verne's reading of Maury's work. See Peter H. Kylstra and Arend Meerburg, "Jules Verne, Maury and the Ocean," *Proceedings of the Second International Congress on the History of Oceanography* (Edinburgh: Royal Society of Edinburgh, 1972), 243–251.

56. Guy T. Houvenaghel, "The First International Conference on Oceanography (Brussels, 1853)," in *Ocean Sciences: Their History and Relation to Man, Proceedings of the 4th International Congress on the History of Oceanography*, ed. Walter Lenz and Margaret Deacon (Hamburg, Germany: Deutsche Hydrographische Zeitschrift, 1990), 335.

57. Maury, *Physical Geography of the Sea*, xiii.

58. See Penelope Hardy, "Every Ship a Floating Observatory: Matthew Fontaine Maury and the Acquisition of Knowledge at Sea," in *Soundings and Crossings: Doing Science at Sea, 1800–1970*, ed. Katherine Anderson and Helen Rozwadowski (Sagamore Beach, MA: Science History, 2016), 19–48.

59. For establishment of the Fish Commission, see Hardy, "Every Ship a Floating Observatory," 67.

60. Kathleen Broome Williams, "From Civilian Planktonologist to Navy Oceanographer: Mary Sears in World War II," in *The Machine in Neptune's Garden: Historical Perspectives on Technology and the Marine Environment*, ed. Helen Rozwadowski and David K. van Keuren (Sagamore Beach, MA: Science History, 2004), 253.

61. Harold Burstyn argues, however, that "as a key sector of the American economy" the maritime industry "provided the economic base on which to develop viable scientific institutions." Institutions like the Coast Survey, the Depot of Charts and Instruments, and the Nautical Almanac Office were important to the broader development of an American scientific community. Harold Burstyn, "Seafaring and the Emergence of American Science," in *The Atlantic World of Robert G. Albion*, ed. Benjamin W. Labaree (Middleton, CT: Wesleyan University Press, 1975), 77.

62. See Alain Corbin, *The Lure of the Sea: The Discovery of the Seaside, 1750–1840* (London: Penguin, 1995).

63. Philip E. Steinberg, *The Social Construction of the Ocean* (Cambridge: Cambridge University Press, 2001), 119.

64. Natascha Adamowsky writes, "[Aquaria] promised a view beyond standing borders, into the deep.... One might behold one of the last great realms of mystery on earth: the oceanic depths and all they concealed." Natascha Adamowsky, *The Mysterious Science of the Sea, 1775–1943* (London: Routledge, 2016), 103.

65. Shirley Hibberd, *Rustic Adornments for Homes of Taste* (London: Groombridge & Sons, 1856), 11.

66. These texts were one branch of a growing body of work that aimed to popularize science for a broad readership. See Bernard Lightman, *Victorian Popularizers of Science: Designing Nature for New Audiences* (Chicago: University of Chicago Press, 2007).

67. Bernd Brunner, *The Ocean at Home: An Illustrated History of the Aquarium* (New York: Princeton Architectural Press, 2005), 100.

68. Sarah Gooll Putnam diary 2, 27 June 1861 to 19 March 1862, Massachusetts Historical Society, Collections Online, http://www.masshist.org/database/viewer.php?item_id=2308.
69. Amanda Bosworth, "Barnum's Whales: The Showman and the Forging of Modern Animal Captivity," *Perspectives on History*, April 2018, https://www.historians.org/publications-and-directories/perspectives-on-history/april-2018/barnums-whales-the-showman-and-the-forging-of-modern-animal-captivity.
70. Henry D. Butler, *The Family Aquarium; or, Aqua Vivarium* (New York: Dick & Fitzgerald, 1858), 5.
71. Richard W. Flint, "American Showmen and European Dealers: Commerce in Wild Animals in Nineteenth-Century America," in *New Worlds, New Animals: From Menagerie to Zoological Park in the Nineteenth Century*, ed. R. J. Hoage and William A. Deiss (Baltimore, MD: Johns Hopkins University Press, 1996), 104.
72. Phineas T. Barnum, *Struggles and Triumphs: Or, Forty Years' Recollections of P. T. Barnum* (Buffalo, NY: Warren, Johnson, 1872), 414.
73. Jamie L. Jones, "Fish out of Water: The 'Prince of Whales' Sideshow and the Environmental Humanities," *Configurations* 25, no. 2 (Spring 2017): 189–214.
74. Hibberd, *Rustic Adornments for Homes of Taste*, 12.
75. Butler, *Family Aquarium*, 11.
76. Philip Henry Gosse, *A Naturalist's Rambles on the Devonshire Coast* (London: John Van Voorst, Paternoster Row, 1853), 47.
77. Gosse, 233–234.
78. Philip Henry Gosse, *The Aquarium: An Unveiling of the Wonders of the Deep Sea* (London: John Van Voorst, Paternoster Row, 1857), 25.
79. Gosse, 276.
80. Gosse, 261. Gosse also experimented with the manufacture of artificial saltwater; Philip Henry Gosse, "On Manufactured Sea-Water for the Aquarium," *Annals and Magazine of Natural History*, July 1854.
81. Philip F. Rehbock, "The Victorian Aquarium in Ecological and Social Perspective," in *Oceanography: The Past*, ed. Mary Sears and Daniel Merriman (New York: Springer-Verlag, 1980), 533.
82. Herdman, *Founders of Oceanography*, 39.
83. For dredging with Forbes, see Daniel Merriman, "Edward Forbes—Manxman," *Progress in Oceanography* 3 (1965), 202.
84. Deacon, *Scientists and the Sea*, 306.
85. Alphonse Milne-Edwards, "Observations sur l'existence de divers mollusques et zoophytes a de très grandes profondeurs dans la mer Méditerranée," *Annales des Sciences Naturelle, zoologie* 4, no. 15 (1861): 149–157.
86. See Eric Mills, "H.M.S. Challenger, Halifax, and the Reverend Dr. Honeyman," *Dalhousie Review* 53, no. 3 (1973): 533.
87. Maurice Yonge, "The Inception and Significance of the *Challenger* Expedition," *Proceedings of the Royal Society of Edinburgh*, section B, 72 (1972): 10.
88. Gary Weir, *An Ocean in Common: American Naval Officers, Scientists, and the Ocean Environment* (College Station: Texas A&M University Press, 2001), 3. John Murray,

a member of *Challenger*'s scientific corps, wrote, "[*Challenger*] circumnavigated the world, traversed the great oceans in many directions, made observations in nearly all departments of the physical and biological sciences, and laid down the broad general foundations of the recent science of oceanography." Johan Hjort and John Murray, *The Depths of the Ocean: A General Account of the Modern Science of Oceanography Based Largely on the Scientific Researches of the Norwegian Steamer Michael Sars in the North Atlantic* (London: Macmillan, 1912), 11.

89. See Antony Adler, "The Ship as Laboratory: Making Space for Field Science at Sea," *Journal of the History of Biology* 47, no. 3 (2014): 333–362.

90. See, for example, Eric Linklater, *The Voyage of the Challenger* (New York: Doubleday, 1972); Richard Corfield, *The Silent Landscape: The Scientific Voyage of HMS Challenger* (Washington, DC: Joseph Henry, 2003). For some cautionary remarks on Corfield's work, see Eric Mills, review of *The Silent Landscape: The Scientific Voyage of HMS Challenger*, by Richard Corfield, *Limnology and Oceanography Bulletin* 13, no. 1 (2004): 4–6.

91. Murray described the dredge used aboard the *Challenger* as differing "little from the ordinary beam trawl of fishermen." John Murray, "The Cruise of the Challenger: First Lecture Delivered in the Hulme Town Hall, Manchester, December 11th, 1877," in *Science Lectures for the People: Science Lectures Delivered in Manchester*, Ninth series (Manchester: John Heywood, 1877), 109.

92. Rozwadowski, *Fathoming the Ocean*, 167–168.

93. Margaret Deacon and Colin Summerhayes write, "One of the paradoxes of the *Challenger* Expedition was that it actually represented the culmination of a process that was serving to turn the attention of scientists back towards the study of the ocean." Margaret Deacon and Colin Summerhayes, introduction to *Understanding the Oceans: A Century of Ocean Exploration*, ed. Margaret Deacon, Tony Rice, and Colin Summerhayes (London: Routledge, 2000), 5. This is also in keeping with Harold Burstyn's appraisal of the *Challenger* expedition and subsequent report as "the major nineteenth-century example of 'big science.'" Harold Burstyn, "Science and Government in the Nineteenth Century: The *Challenger* Expedition and Its Report," *Bulletin de l'Institut Océanographique*, no. 2 (1968): 606.

94. Thomson, *The Voyage of the Challenger: The Atlantic—a Preliminary Account* [. . .], Vol. 1 (London: Macmillan and Co., 1877), 51.

95. Murray, *Cruise of the Challenger*, 105.

96. "The collections, when finally assembled at Sheerness, after the return of the ship, were contained in 2,270 jars, 1,749 bottles, 1,860 glass tubes, and 176 tin cases of alcoholics, with 22 casks of specimens in brine, and 180 tin cases of dried specimens, besides large quantities of material already sent home from Bermuda, Halifax, Cape Town, Sydney, Hong Kong, and Japan." "The Exploring Voyage of the Challenger," *Science* 3, no. 66 (9 May 1884): 576.

97. Daniel Merriman, "Challengers of Neptune: The 'Philosophers.'" *Proceedings of the Royal Society of Edinburgh*, section B, 72 (1972): 19.

98. "Deep Sea Searches," *New York Daily Herald*, 19 May 1873, 5.

99. Henry H. Higgins, "The 'Challenger' Collections," *Nature*, 25 January 1877, 274.

100. Deacon, *Scientists and the Sea*, 367.

101. In an 1877 letter to Thomson, claiming a precedent for this distribution, and implying the justice of reciprocity, Agassiz wrote, "As far back as 1869 when Porcupine had just returned we had agreed upon the wisdom of letting the same people work up all the deep sea things from both sides of the Atlantic, as far as practicable." Deacon, 394.
102. For the destination of the sponges, see Deacon, 367. For the Crustacea, see Hubbard, *Science on the Scales*, 21.
103. Alexander Agassiz, "Zoology of the 'Challenger' Expedition," *Annals and Magazine of Natural History* 19, no. 111 (1877): 276.
104. P. Martin Duncan, "Miscellaneous. Zoology of the 'Challenger' Expedition. To the Editors of the Annals and Magazine of Natural History," *Annals and Magazine of Natural History* 19, no. 113 (1877): 429–430.
105. Duncan, 429.
106. "The 'Challenger' Collections," *Nature*, 14 June 1877, 118.
107. It is worth noting the presence of a German naturalist as one of the scientific staff aboard the *Challenger*—R. Von Willemoes-Suhm—who succumbed to illness in the course of the journey. His participation was happenstance, the result of a last-minute invitation by Thomson. See Charles Wyville Thomson, "R. Von Willemoes-Suhm," *Nature*, 2 December 1875, 88–89.
108. Rozwadowski, *Fathoming the Ocean*, 173.
109. See Eric Mills, "One 'Different Kind of Gentleman': Alfred Merle Norman (1831–1918), Invertebrate Zoologist," *Zoological Journal of the Linnaean Society* 68 (January 1980): 84–85.
110. Alexander Agassiz to John Murray, 10 March 1882. University of Edinburgh Special Collections.
111. Harold L. Burstyn, "Science Pays Off: Sir John Murray and the Christmas Island Phosphate Industry, 1886–1914," *Social Studies of Science* 5 (1975): 5–6. Burstyn argues that the cost to the government was recouped through the *Challenger*'s discovery of phosphate deposits on Christmas Island in the Indian Ocean.
112. Harold Burstyn, "'Big Science' in Victorian Britain," in Deacon, Rice, and Summerhayes, *Understanding the Oceans*, 51.
113. Deacon, *Scientists and the Sea*, 369.
114. Hubbard, *Science on the Scales*, 150.
115. In France, the sardine fishery was particularly important to the economy in Brittany and during the early twentieth century periodically collapsed. Tim D. Smith, *Scaling Fisheries: The Science of Measuring the Effects of Fishing, 1855–1955* (Cambridge: Cambridge University Press, 1994), 14. Olivier Levasseur has called Napoleon III's Third Empire the "golden age" of French aquaculture. Olivier Levasseur, "La culture des mers en France au XIXe siècle," in *Observation des écosystèmes marin et terrestre de la Côte d'Opale: du naturalisme à l'écologie*, ed. François G. Schmitt (Paris: Union des océanographes de France, 2011), 97.
116. "Punch's Sentiments," *Punch Magazine* 7 (1844): 213.
117. "Exposition internationale de pêche et d'aquiculture à Arcachon, en juillet 1886. Formulaire de questions." *Bulletin de la Société Impériale Zoologique d'Acclimatation*

(Paris: Victor Masson et Fils, 1866), 182–188. "[H]ave dominion over the fish of the sea" is from Gen. 1:28 (King James version).

118. "The Exposition of Arcachon and Its Object," *Fraser's Magazine: Fishing Excerpts* 4 (1861–1868): 299.

119. "Rapport à L'Empereur sur l'exposition internationale de Pêche et d'Aquiculture d'Arcachon" (Paris: Typographie E. Panckoucke, 1867). The colonial exposition in Paris in 1900 exhibited the work of the Oceanographic Institute of Indochina. The 1906 colonial exposition of Marseille featured a "palace of the sea," and the international and maritime exposition of Bordeaux in 1907 included displays featuring the work of French marine zoologists, the work of Albert I, prince of Monaco, and charts produced by the Hydrographic Service and the Oceanographic Society of the Gulf of Gascogne. See Jules Girard, "L'océanographie à l'Exposition maritime de Bordeaux," *La Géographie: Bulletin de la Société de Géographie* 16 (1907): 111–113.

120. Frederick Whymper, *The Fisheries of the World: An Illustrated and Descriptive Record of the International Fisheries Exhibition, 1883* (London: Cassell, 1884), 95.

121. Kelly Bushnell, "Monstrosity and Material Culture in Nineteenth-Century English Sea Literature" (PhD diss., Department of English, Royal Holloway, University of London, 2016), 234. There is no published study of the 1883 International Fisheries Exposition; however, it is discussed in detail by Kelly Bushnell. See also the discussion of the Canadian contribution to the exposition in William Knight, "Modeling Authority at the Canadian Fisheries Museum, 1884–1918" (PhD diss., Department of History, Carleton University, 2014).

122. Whymper, *Fisheries of the World*, 96. The US government had previously mounted fisheries exhibits in the United States. The 1867 Centennial Exposition in Philadelphia featured 408 painted plaster models of American fish, organized by the Smithsonian curators Spencer F. Bair and George Brown Goode. Bruno Giberti, *Designing the Centennial* (Lexington: University Press of Kentucky, 2015), 145–146.

123. Bushnell, "Monstrosity and Material Culture," 240.

124. Bushnell, 242.

125. Whymper, *Fisheries of the World*, 107.

126. "The International Fisheries Exhibition.—Second Paper," *Science* 1, no. 20 (22 June 1883): 565.

127. See Xavier Dubois, *La révolution sardinière: pêcheurs et conserveurs en Bretagne Sud au XIXe siècle* (Rennes: Presses universitaires de Rennes, 2004); and Marc Pavé, *La pêche côtière en France (1715–1850): approche sociale et environnementale* (Paris: L'Harmattan, 2013).

128. The company seems to have been based in France, although a Scotsman named W. S. Johnston was the founder. These steam trawlers were built by J. Elder and Company in Scotland. Henry Wood, "Fisheries of the United Kingdom," in *Sea Fisheries: Their Investigation in the United Kingdom*, ed. Michael Graham (London: Edward Arnold, 1956), 16. For more on Elder's Fairfield shipbuilding company, see W. J. Macquorn Rankine, *A Memoir of John Elder Engineer and Shipbuilder* (Edinburgh: William Blackwood & Sons, 1871).

129. *Annuaire Statistique de la France* (Paris: Imprimerie Nationale, 1882), 366.
130. Jeffrey Bolster, *The Mortal Sea: Fishing the Atlantic in the Age of Sail* (Cambridge, MA: Harvard University Press, 2012), 10.
131. Bottom trawl nets had been in use in the waters off the British Isles as far back as the fourteenth century. Ruth Thurstan, Julie Hawkins, and Callum Roberts, "Origins of the Bottom Trawling Controversy in the British Isles: 19th Century Witness Testimonies Reveal Evidence of Early Fishery Declines," *Fish and Fisheries* 15, no. 3 (2013): 2.
132. Julien Thoulet, *A Voyage to Newfoundland*, trans. and ed. Scott Jamieson (Montreal: University of McGill, Queens's University Press, 2005), 50.
133. A later commentator, writing in 1902, compared steam trawling in the North Atlantic fisheries to the decimation of the American buffalo: "When the modern steam trawler, knowing the season when fish crowd into very limited haunts, gets at these grounds with his persistence and effective gear, it is not a too prejudiced view to take when we say that an exterminating hunt has begun." W. S. Green, "The Sea Fisheries of Ireland," in *Ireland: Industrial and Agricultural* (Dublin: Browne & Nolan, 1902), 374.
134. Hubbard, *Science on the Scales*, 150.
135. "Report of the Commissioners Appointed to Inquire into the Sea Fisheries of the United Kingdom, Vol. 1" in *Reports from Commissioners: Twenty-Two Volumes*, Vol. 17 (London: George Edward Eyre and William Spottiswoode, 1866), xvii–xviii.
136. Frank Buckland, *The Natural History of British Fishes; Their Structure, Economic Uses, and Capture by Net and Rod* (London: Gresham, 1881), vii. Geoffrey Burgess writes of Buckland that he "remains the only one consistently calling for research into fishery problems, publicizing the activities of the industry, drawing attention to the national importance of fish in the diet, and acting as a focus for those in the industry and elsewhere who were interested and concerned about its proper commercial development." Geoffrey Burgess, "Frank Buckland and the Buckland Foundation," in *British Marine Science and Meteorology: The History of Their Development and Application to Marine Fishing Problems*, Buckland Occasional Papers, No. 2 (Lowestoft, UK: Buckland Foundation, 1996), 20. Buckland established the first fisheries museum in the United Kingdom in 1865—the Museum of Economic Fish Culture—which he "intended as an Educational means of informing the public, not only as to the Natural History of Fish, but also as to their commercial uses, and as to the development of the fisheries of this country." G. H. O. Burgess, *The Eccentric Ark: The Curious World of Frank Buckland* (New York: Horizon, 1967), 125. Burgess notes that although the museum became a famous entertainment venue, visited by Queen Victoria and other members of the royal family, the information presented was of little scientific value.
137. The marine conservation biologist and writer Callum Roberts has attributed the commission's inability to recognize the destruction of British fisheries to the fact that "in the absence of systematic fishery statistics . . . they were unable to see any clear trend in abundance." Callum Roberts, *The Unnatural History of the Sea* (Washington, DC: Island Press, 2007), 142. As Jennifer Hubbard notes, "Fisheries biology was essentially a branch of science born out of the growing recognition

that human activities were altering the environment." Hubbard, *Science on the Scales*, 149.
138. Thomas Henry Huxley, "Inaugural Address by Professor Huxley," *The Fisheries Exhibition Literature* (London: William Clowes & Sons, 1884), 4:8.
139. Huxley, 14. For lengthier discussion of Huxley's aversion to marine fisheries regulation and his disagreement with Lankester, see Hubbard, *Science on the Scales*, 150–156.
140. Bolster, *Mortal Sea*, 122.
141. E. Ray Lankester, *The Scientific Results of the Exhibition* (London: William Clowes & Sons, 1883), 11.
142. Lankester, 14.
143. David W. Sims and Alan J. Southward, "Dwindling Fish Numbers Already of Concern in 1883," *Nature* 439, no. 9 (February 2006): 660. It has been suggested that Lankester provided the basic layout design for the Plymouth station, the largest of the British stations. A. J. Southward, "The Marine Biological Association and Fishery Research, 1884–1924: Scientific and Political Conflicts That Changed the Course of Marine Science in the United Kingdom," in *British Marine Science and Meteorology*, 67.
144. Lankester, *Scientific Results of the Exhibition*, 21–22.
145. E. Ray Lankester, "An American Sea-Side Laboratory," *Nature*, 25 March 1880, 498.
146. M. J. Pérard and M. Maire, *Congrès International d'Aquiculture et de Pêche Tenue à Paris du 14 au 19 Septembre 1900. Procès—Verbaux Sommaires* (Paris: Imprimerie Nationale, 1900), 13.
147. Quoted in Rozwadowski, *Fathoming the Ocean*, 179.
148. As a scientific instrument, plankton nets were of equal importance to the development of biological oceanography. John Murray attempted to gather deepwater plankton during the *Challenger* expedition using muslin tow nets. See Eric Mills, "Alexander Agassiz, Carl Chun and the Problem of the Intermediate Fauna," in *Oceanography: The Past*, ed. Mary Sears and Daniel Merriman (New York: Springer-Verlag, 1980), 362; and Peter H. Wiebe and Mark C. Benfield, "From the Hensen Net toward Four-Dimensional Biological Oceanography," *Progress in Oceanography* 56 (2003): 7–136.

2. Marine Science for the Nation or for the World?

Epigraphs: Albert I, prince of Monaco, *La carrière d'un navigateur*, 2nd ed. (Monaco: Imprimerie de Monaco, 1905), 34; Arthur Schuster, "International Science," *The University Review* 3 (1906): 262.

1. Eric Mills, "Problems of Deep-Sea Biology: An Historical Perspective," in *The Sea: Ideas and Observations on Progress in the Study of the Seas*, ed. Gilbert T. Rowe (John Wiley & Sons, 1983), 47.
2. Louis Barriety, "Océanographie du Golfe de Gascogne: Un précurseur: le marquis Léopold de Folin," in *Comptes rendus du 94ᵉ congrès national des sociétés savants* (Paris: Bibliothèque Nationale, 1969), 1:131–140.

3. Helen Rozwadowski, *Fathoming the Ocean: The Discovery and Exploration of the Deep Sea* (Cambridge, MA: Harvard University Press, 2005), 173.
4. These expeditions did help launch the field of marine microbiology. See Antony Adler and Erik Dücker, "When Pasteurian Science Went to Sea: The Birth of Marine Microbiology," *Journal of the History of Biology* 51, no. 1 (March 2018): 107–133.
5. There were efforts to promote oceanography at the local level. The Paris-based Société de Géographie formed a commission in 1874 and charged it with developing a program for oceanographic observations carried out by naval and civilian vessels, and in 1899 a Société d'Océanographie du Golfe de Gascogne formed in Bordeaux. Jacqueline Carpine-Lancre, "Les Expéditions Océanographiques Françaises du XIXe Siècle," *Actes de 12e Congrès international d'histoire des sciences*, Vol. 7 (Paris: A. Blanchard, Librairie Scientifique et Technique, 1971), 62.
6. See Homer A. Jack, "Biological Field Stations of the World" *Chronica Botanica* 9, no. 1 (1945), 1–73.
7. William Herdman, *Founders of Oceanography and Their Work: An Introduction to the Science of the Sea* (London: Edward Arnold, 1923), 135.
8. As Raf De Bont explains, "Field stations provide researchers with their own scientific habitat, within the larger habitat of the organisms they study." Raf De Bont, *Stations in the Field: A History of Place-Based Animal Research, 1870–1930* (Chicago: University of Chicago Press, 2015), 3.
9. In her 1988 study of the Naples station and the Marine Biological Laboratory at Woods Hole, Jane Maienschein shows that naturalists turned to the seashore for the variety of organisms obtainable and for the advantages offered by studying organisms in their natural environment. Crucially, she identifies marine stations as repositories of expertise where, "by the 1880s in Naples or the 1890s in Woods Hole, one could obtain expert advice on where to look for organisms or could request them and have them reliably appear at one's lab table." Jane Maienschein, "History of American Marine Laboratories: Why Do Research at the Seashore," *American Zoologist* 28, no. 1 (1988): 22. See also Keith Benson, "Summer Camp, Seaside Station, and Marine Laboratory: Marine Biology and Its Institutional Identity," *Historical Studies in the Physical and Biological Sciences* 32, no. 1, Second Laboratory History Conference (2001): 11–18.
10. Cuvier passed on his inclination for seaside fieldwork collecting to many of his students, most famously to Henri Milne-Edwards (father of Alphonse Milne-Edwards) and Jean-Victor Audouin. Henri Milne-Edwards, one of the first naturalists to use an underwater breathing apparatus for scientific fieldwork, was the mentor of Henri Lacaze-Duthiers. Edward A. Eigen, "Between Stations and Habitations: The Architecture of French Science at the Shore, 1830–1900" (PhD diss., MIT Department of Architecture, 2000), 26–27. See also Jean Théodoridès, "Les débuts de la biologie marine en France: Jean-Victor Audouin et Henri Milne-Edwards, 1826–1829," in *Actes du 1er Congrès International de l'Océanographie (Monaco, 1966)* (Monaco: Musée océanographique de Monaco, 1968), 2:417–437. For Cuvier's coastal work, see Charles Couston Gillispie, *Science and Polity in France: The End of the Old Regime*, (Princeton, NJ: Princeton University Press, 1980), 452.

11. See Antony Adler, "The Hybrid Shore: The Marine Station Movement and Scientific Uses of the Littoral, 1843–1910," in *Soundings and Crossings: Doing Science at Sea, 1800–1970*, ed. Katharine Anderson and Helen M. Rozwadowski (Sagamore Beach, MA: Science History Publications, 2016).
12. Charles Kofoid, *The Biological Stations of Europe* (Washington, DC: Government Printing Office, 1910), 35.
13. Victor Coste, "De l'observation et de l'expérience en physiologie," *Journal de l'anatomie et de la physiologie normales et Pathologiques de l'homme et des animaux* 6 (1869): 659. Coste enjoyed the patronage of Napoleon III, having previously been charged with writing a government survey report on fisheries and aquiculture in France and Italy. Victor Coste, *Voyage d'exploration sur les côtes de France et d'Italie* (Paris: Imprimerie Impériale, 1855).
14. "Pisciculture—l'établissement de Concarneau," *Magasin Pittoresque* 37 (1869): 300. Henri Lacaze-Duthiers, "Les Laboratoires Maritimes de Roscoff et de Banyuls en 1891," *Archives de Zoologie Expérimentale et Générale*, série 2, tome 9 (1891): 256.
15. René Sand, "Les Laboratoires Maritimes de Zoologie," Extrait de la *Revue de l'Université de Bruxelles*, tome 3 (1897–1898) (Brussels: Bruylant-Christophe & Cie, 1897), 10.
16. The Algerian "coral fishery" was a valuable commercial export industry of the French colony. For a lengthier description of this period of Lacaze-Duthiers's life and his voyages to the Atlantic and Mediterranean coasts, see Henri Lacaze-Duthiers, *Notice sur les travaux scientifiques de M. F.—J.—Henri de Lacaze-Duthiers présentée a l'appui de sa candidature à l'académie des sciences* (Paris: Institut de France, 1862).
17. Henri de Parville, "Henri de Lacaze-Duthiers," *La Nature*, no. 1471 (3 August 1901). When Charles Darwin learned that Lacaze-Duthiers would vote in support of his election as a corresponding member of the French Academy in 1872, he wrote that he was delighted, having "long honored his name." Charles Darwin, *The Life and Letters of Charles Darwin Including an Autobiographical Chapter*, ed. Francis Darwin (London: John Murray, 1887), 3:155.
18. Louis Joubin, "Le Laboratoire Zoologique de Roscoff," *La Nature*, no. 648 (31 October 1885). I have relied on the summaries provided by the American oceanographer Charles Kofoid, who visited both stations while compiling a report on the biological stations of Europe in 1910. He cites Lacaze-Duthiers's reports in the *Archives de zoologie expérimentales*. For a detailed history of the Roscoff station, see Josquin Debaz, "Une histoire de la station de biology marine de Roscoff (1872–1914)" (unpublished manuscript), https://halshs.archives-ouvertes.fr/halshs-00380634/document. See also Josquin Debaz, "Les stations françaises de biologie marine et leurs périodiques entre 1872 et 1914" (PhD diss., École des Hautes Études en Sciences Sociales, Centre Alexandre Koyré, 2005). For Charles Kofoid and his study of European stations, see Deborah Day and Eric L. Mills, "Charles Atwood Kofoid and the Biological Stations of Europe," in *Places, People, Tools: Oceanography in the Mediterranean and Beyond. Proceedings of the Eighth International Congress for the History of Oceanography*, ed. Christiane Groeben (Naples, Italy: Giannini), 233–254.

19. Henri Lacaze-Duthiers, "A propos de la station des choetoptères et des myxicoles sur les plages de Roscoff et de Saint-Pol-De-Léon, côtes de Bretagne (finistère)," *Archives de Zoologie Expérimentale et Générale: Notes et Revue*, tome 1 (1872): xvii. As discussed in Chapter 1, in 1839 the British Association for the Advancement of Science had formed a "dredging committee," which led a coordinated effort to investigate marine life in the seas surrounding Great Britain. Philip Rehbock, "The Early Dredgers: 'Naturalizing' in British Seas, 1830–1850," *Journal of the History of Biology* 12, no. 2 (1979): 328.
20. The first major report of the dredging committee was produced in 1850. See Rehbock, "The Early Dredgers," 341.
21. Rehbock, 341.
22. Initially, Lacaze-Duthiers and his students worked out of a hotel; a local woman was hired to bring a fresh jug of salt water each day to keep alive the organisms they studied in household containers and soup bowls adapted to serve as aquaria. Edmond Perrier, "Précédé," Armand Quatrefages, *Les émules de Darwin* (Paris: Ancienne Librairie Gemer Bailliere, Félix Alcan, 1894), 1:xiii.
23. Harry Paul, *From Knowledge to Power: The Rise of the Science Empire in France, 1860–1939* (Cambridge: Cambridge University Press, 1985), 108.
24. Jean Chalon, "Quelques mots sur Roscoff," *Bulletins de la Société royale de botanique de Belgique* 37 (1898): 107.
25. Eugen Weber, *Peasants into Frenchmen: The Modernization of Rural France, 1870–1914* (Palo Alto, CA: Stanford University Press, 1976), 486.
26. Weber, 4–5.
27. Weber, 100.
28. Henri Lacaze-Duthiers, "Notes et Revues," *Archives de zoologie expérimentale et générale*, tome 1 (Paris: Librarie Germer Bailliere, 1872), xix.
29. Henri Lacaze-Duthiers, "Avertissement," *Archives de zoologie expérimentale et générale*, tome 1 (Paris: Librarie Germer Bailliere, 1872), v–vi.
30. Harry Paul describes Lacaze-Duthiers's support for the popularization of science as adhering to "one of the basic republican dogmas." Paul, *From Knowledge to Power*, 111.
31. "Progress of a French Zoological Station," *Times* (London), 3 March 1881.
32. Lacaze-Duthiers felt he could declare Roscoff and Banyuls completed in 1891, noting that gradual improvement would always be necessary but that the "means of research" for both stations "were largely assured." Lacaze-Duthiers, "Les Laboratoires Maritimes de Roscoff et de Banyuls en 1891," 254.
33. Henri Lacaze-Duthiers, "Le monde de la mer at ses laboratoires," *Revue scientifique* 42 (1888): 37.
34. Lacaze-Duthiers, 37. Lacaze-Duthiers's critique is misleading because the contract for the visiting research positions at the Naples station explicitly gave permission for attendees to participate in fishing and to learn the methods of collection. See C. W. Stiles, "Report on the Memorial Presented to the Smithsonian Institution Regarding an American Table at the Naples Zoological Station," *Science* 21, no. 541 (16 June 1893): 329.

35. Henri Lacaze-Duthiers, "Les progrès de la station zoologique de Roscoff et la création du laboratoire Arago à Banyuls-sur-mer," *Archives de Zoologie Expérimentale et Générale*, tome 9 (1881): 564.
36. Maurice Fontaine, "From the Physiology of Marine Organisms to Oceanographic Physiology or Physiological Oceanography," in *Oceanography: The Past*, ed. Mary Sears and Daniel Merriman (New York: Springer-Verlag, 1980), 354. Frédéricq's work led to the discovery of hemocyanins.
37. See Richard Burkhardt Jr., "Naturalists' Practices and Nature's Empire," *Pacific Science* 55, no. 4 (2001): 327–341.
38. Raf De Bont, "Between the Laboratory and the Deep Blue Sea: Space Issues in the Marine Stations of Naples and Wimereux," *Social Studies of Science* 39, no. 2 (2009): 215. See also Michel Glémarec, *Qu'est-ce que la biologie marine? De la biologie marine à l'océanographie biologique* (Paris: Vuibert, 2010), 72–75. Tatihou received approximately 25 percent of its funding from the Ministry of Fisheries. Tim D. Smith, *Scaling Fisheries: The Science of Measuring the Effects of Fishing, 1855–1955* (Cambridge: Cambridge University Press, 1994), 26.
39. Roscoff visitor log entry, 24 August 1888. Roscoff Station Archives. Unidentifiable signature.
40. Roscoff visitor log entry, 17 July 1888. Roscoff Station Archives. This entry may have been written by the Scottish biologist Sir D'Arcy Wentworth Thompson, but the signature is difficult to decipher.
41. Lacaze-Duthiers, "Le monde de la mer," 39.
42. Mills has termed this "the paradox of French marine science: the virtual invisibility of France at a time when dynamic oceanography was making inroads into the way that physical oceanography was being practiced in Scandinavia, Germany, and the United States." Eric Mills, *The Fluid Envelope of Our Planet: How the Study of the Ocean Currents Became a Science* (Toronto: University of Toronto Press, 2009), 163.
43. The French government entered a contract to rent tables with the Naples station finally in 1919. Jean-Louis Fischer, "L'aspect social et politique des relations épistolaires entre quelques savants français et la Station zoologique de Naples de 1878 à 1912," *Revue d'histoire des sciences* 33, no. 3 (1980): 228. The French government was not alone in balking at the cost. The American oceanographer Alexander Agassiz, president of the Harvard Museum of Comparative Zoology, also complained that the cost of maintaining three or four students at Naples was "a serious expenditure for any university, and out of proportion to the expenses incurred for them in other departments." Alexander Agassiz, *Annual Report of the Museum of Comparative Zoology* (Cambridge: Cambridge University Press, 1895), 6. Agassiz notes that the cost of a table at Naples was 100 lira per year.
44. Jane M. Oppenheimer, introduction to *Von Baer–Dohrn Correspondence* (Philadelphia: American Philosophical Society, 1993), 4.
45. Christiane Groeben, "Anton Dohrn: The Statesman of Darwinism," *Biological Bulletin* 168 (June 1985): 5.
46. As Lankester later recalled, "He was what appeared to me, with my English upbringing, singularly introspective, and he puzzled, even occasionally alarmed,

me by his self-conscious and systematic cultivation of his will-power." E. Ray Lankester, "Anton Dohrn," *Nature*, 7 October 1909, 430.
47. In the Strait of Messina tidal currents cause upwelling. It has long been a favored collecting area for naturalists because deep-sea fauna are often brought to the surface.
48. Dohrn to Darwin, 30 December 1869, Darwin Correspondence Project, "Letter no. 7038," accessed on 3 April 2019. https://www.darwinproject.ac.uk/letter/DCP-LETT-7038.xml.
49. De Bont, *Stations in the Field*, 59.
50. Maienschein, "History of American Marine Laboratories," 18.
51. De Bont, "Between the Laboratory and the Deep Blue Sea," 209. See also Christiane Groeben, "Tourists in Science: 19th Century Research Trips to the Mediterranean," suppl. 1, *Proceedings of the California Academy of Sciences—Fourth Series* 59, no. 9 (September 2008): 139–154.
52. De Bont, "Between the Laboratory and the Deep Blue Sea," 210.
53. Locals in Roscoff opposed the expansion of the laboratory buildings. There was also conflict over access to the tidal flats. For naturalists this was a place for gathering scientific specimens; for the people of Roscoff, a place for gathering food and fertilizer for their fields.
54. Dohrn to Darwin, 15 February 1872, Dohrn Darwin Correspondence, 39.
55. Theodor Heuss, *Anton Dohrn: A Life for Science* (Berlin: Springer-Verlag, 1991), 151.
56. Keith Benson, "Review Paper: The Naples Stazione Zoologica and Its Impact on the Emergence of American Marine Biology," *Journal of the History of Biology* 21, no. 2 (1988): 333.
57. Kofoid, *Biological Stations of Europe*, 8.
58. Kofoid, 7.
59. Nansen recorded his impressions of the Naples station in an article published in 1887 in the Norwegian journal *Naturen*. I have relied on excerpts provided in the English-language translation of Nansen's biography. Waldemar Christopher Brøgger and Nordahl Rolfsen, *Fridtjof Nansen, 1861–1893*, trans. William Archer (London: Longmans, Green, 1896), 109.
60. Christiane Groeben, "The Stazione Zoologica: A Clearing House for Marine Organisms," in *Oceanographic History: The Pacific and Beyond*, ed. Keith Benson and Philip Rehbock (Seattle: University of Washington Press, 2002), 542.
61. Jennifer Hubbard, *A Science on the Scales: The Rise of Canadian Atlantic Fisheries Biology, 1898–1939* (Toronto: University of Toronto Press, 2006), 35.
62. Much of the work carried out at Naples and the Marine Biological Laboratories in Woods Hole, Massachusetts, was aimed at understanding biological processes rather than at tackling questions related specifically to marine ecology. Paul Farber, *Finding Order in Nature: The Naturalist Tradition from Linnaeus to E. O. Wilson* (Baltimore: Johns Hopkins University Press, 2000), 81. Urchins are useful model organisms because they can be artificially stimulated to spawn. A single urchin produces tens of thousands of eggs. Furthermore, sea urchin cells, like those of all echinoderms, undergo radial holoblastic cleavage, which facilitates comparative study.

63. Dohrn to Von Baer, 8 February 1873, *Correspondence: Karl Ernst von Baer–Anton Dohrn*, ed. Christiane Groeben (Philadelphia: American Philosophical Society, 1993), 46.
64. In Liselotte Dieckman's translation of the quotation provided by Theodor Heuss, Dohrn refers to Lacaze-Duthier as "my special competitor" in the letter to his sister. Heuss, *Anton Dohrn: A Life for Science*, 239.
65. See M. M. Caullery, "La station zoologique de Naples. Pourquoi la France y doit avoir une place," *Revue scientifique* 6, no. 25 (December 1906): 779, 782.
66. Henri Lacaze-Duthiers, *Rapport sur l'École practique des hautes études 1877–1878, 1878–1879* (Paris: Imprimerie nationale, 1879), 90. Translated in Selim A. Morcos, "Marine Sciences in Egypt," in *Places, People, Tools: Oceanography in the Mediterranean and Beyond—Proceedings of the Eight International Congress for the History of Oceanography*, ed. Christiane Groeben (Naples, Italy: Giannini, 2013), 19. Morcos notes that Egypt had historically been a site of competing colonial interests between Britain and France.
67. Norway and Sweden also signed the convention but did not ratify it. Rozwadowski, *The Sea Knows No Boundaries: A Century of Marine Science under ICES* (Copenhagen: International Council for the Exploration of the Sea, 2002), 13.
68. Mills, *Fluid Envelope of Our Planet*, 84.
69. Among the scientific participants were Martin Knudsen (1871–1949) of Denmark, Fridtjof Nansen (1861–1930) of Norway, and Otto Pettersson (1848–1941) of Sweden. Jens Smed, "Physical Oceanography in Scandinavia," *Earth-Science Reviews: International Magazine for Geo-Scientists* 3 (1967): A119–A121.
70. Julien Thoulet maintained correspondence with the Swedish marine chemist and ICES founder Otto Pettersson. In the Thoulet-Pettersson correspondence is also repeated reference to Thoulet's collaboration and correspondence with the Danish physicist Martin Knudsen (1871–1949). In a letter sent in 1889, Thoulet informed Pettersson that the previous year he had been charged with a mission—he does not say by whom, but presumably by the French state—to travel to Christiana and Edinburgh to study the oceanographic work conducted in Norway and Scotland. Thoulet also requested several of Pettersson's publications. Julien Thoulet to Otto Pettersson, 16 May 1889, photocopy of letter in the archives of the University of Gothenburg, from the collection of Jacqueline Carpine-Lancre.
71. For some discussion on France's reluctance to join ICES, see Jens Smed, "The Accession of France into ICES," *History of Oceanography* 21 (September 2009): 9–17.
72. France finally joined ICES in 1914, and further delays due to World War I meant that France became a full member only in 1921.
73. Thoulet to Pettersson, 11 March 1903, photocopy of letter in the archives of the University of Gothenburg, from the collection of Jacqueline Carpine-Lancre.
74. In 1911 he was appointed to the Conseil Supérieur des Pêches Maritimes, organized by the Ministry of the Marine.
75. Julien Thoulet, *L'Océan: Ses Lois et Ses Problèmes* (Paris: Librairie Hachette, 1904), 60–62.
76. Thoulet, 217.

77. Thoulet to Pettersson, 9 May 1903, photocopy of letter in the archives of the University of Gothenburg, from the collection of Jacqueline Carpine-Lancre. Thoulet did receive support from the French Navy but by 1889 he was aware of the limits of that aid. His first expedition was nevertheless aboard a naval vessel; the navy covered the cost of his first treatise on oceanography (1890), and for many years he worked as an instructor of marine science at the École supérieure de marine. See Camille Vallaux, "Nécrologie—Julien-Olivie Thoulet (1843–1936)," *Annales de Géographie* 45, no. 254 (1936): 217–218.
78. Albert, *La carrière d'un navigateur*, 2nd ed. (Monaco: Imprimerie de Monaco, 1905), 30.
79. "Count Castellane Scores Prince Albert of Monaco," *San Francisco Chronicle*, 5 July 1899.
80. The casino is intended for the use solely of foreigners. Monegasques have since the reign of Charles III been forbidden by law from gambling in the casino.
81. Jacqueline Carpine-Lancre, *Albert I Prince of Monaco (1848–1922)* (Monaco: Editions EGC, 1998), 6.
82. "More Reform in Monaco; Prince Albert Surrenders Control of the Government Finances," *New York Times*, 16 October 1910. See also a very interesting interview with Prince Albert I, published by the *New York Times* in 1922, in which he suggests that his granting of the constitution secured Monaco's independence from an impending German takeover. Gualtiero Campino, "The Paradox of Monaco. How Late Prince Had Fairly to Force Constitution on His Reluctant People," *New York Times*, 2 July 1922.
83. *Albert Ier, Prince de Monaco. Des oeuvres de science, de lumière et de paix. 150e Anniversaire de sa Naissance*, 24. Albert's initial contact with the Paris museum probably dates closer to the 1870s; an ardent hunter, he employed the same taxidermist as the museum. For the publication of the scientific reports associated with his expeditions, Albert relied on the expertise of professors at the museum. Of the sixty-four scientific campaign reports published between 1889 and 1922, twenty-one have a museum professor or staff assistant as author or coauthor. Jacqueline Carpine-Lancre, "Honorer la science: Le Prince Albert Ier de Monaco, le Muséum de Paris et Lamark," *Bulletin du Musée Anthropologique at Préhistorique de Monaco* 51 (2011): 156.
84. He paid tribute to Milne-Edwards in speech at the inauguration of the Oceanographic Museum, referring to Milne-Edwards as his mentor. Albert I, prince of Monaco, "Discours de S.A.S. le Prince de Monaco. Discours prononcés à l'occasion des fêtes d'inauguration du Musée océanographique de Monaco, 29 March 1910," 7, https://www.worldcat.org/title/discours-prononces-a-loccasion-des-fetes-dinauguration-du-musee-oceanographique-de-monaco-29-mars-1910-ler-avril-1910/oclc/28597377.
85. Albert Ier, "Discours à l'inauguration de l'Institut de paléontologie humaine," *Journal de Monaco* 63, no. 3283 (1920). Quoted in Jacqueline Carpine-Lancre, "Le Prince Albert Ier de Monaco et la Science," *Archives de l'Institut de paléontologie humaine*, vol. 39 (2008), 15.

86. Prince Albert had four different yachts, and he alternated between two names: *Hirondelle* and *Princesse Alice*. It has become common practice to add "I" and "II" to differentiate them. Albert provides a description of the *Hirondelle* crew in his autobiography. Albert, *La carrière d'un navigateur*, 48–54.
87. Prince Albert maintained correspondence with the American oceanographer Alexander Agassiz, to whom he was introduced by Alphonse Milne-Edwards in 1898. Agassiz visited Monaco in 1906, and Albert invited him to serve on the scientific advisory board of his Oceanographic Institute. For Prince Albert's use of an American-made current meter, see Christian Carpine, *La pratique de l'océanographie au temps du Prince Albert Ier* (Monaco: Musée Océanographique, 2002), 98. See also John Elliott Pillsbury, "The Grandest and Most Mighty Terrestrial Phenomenon: The Gulf Stream," *National Geographic Magazine* 23, no. 7 (1912): 767–778.
88. Albert I, "A New Ship for the Study of the Sea," *Proceedings of the Royal Society of Edinburgh* 18 (November 1890–July 1891): 298.
89. Jacqueline Carpine-Lancre, *Albert Ier, Prince de Monaco. Des oeuvres de science, de lumière et de paix* (Monaco: Palais de S.A.S. Le Prince, 1998), 24–25.
90. A partial list of visitors who left calling cards at the pavilion is recorded in a report to Albert from Jules Richard. Richard to Albert, 11 September 1889, Palace Archives, Monaco. For a description of some features of the 1900 Monaco exposition in Paris, see Richard to Albert, 19 April 1900, Richard to Albert, 26 April 1900, and Richard to Albert, 29 April 1900, all in the Palace Archives, Monaco. As Carpine-Lancre has argued, Albert's participation in international conferences and expositions (Paris in 1889 and 1900, Saint Louis in 1904, Milan and Marseille in 1906, Brussels in 1910, and Genoa in 1914) amplified the prestige of Monaco on the world stage. Carpine-Lancre, *Albert I Prince of Monaco (1848–1922)*, 10.
91. Albert married the New Orleans–born aristocratic divorcée Marie Alice Heine (1858–1925) on 30 October 1889.
92. A description of some of the *Princesse Alice*'s features is provided in F. Faideau, "Le Yacht 'Princesse Alice,'" *La Science Illustrée* 17 (1896): 245–246.
93. Albert, "A New Ship for the Study of the Sea," 295.
94. Among them were also two women: Jeanne Le Roux, an artist, and Hanna Resvoll-Holmsen, a Norwegian botanist.
95. Buchanan also took part in the 1902 summer campaign of the *Princesse Alice II*. He and Prince Albert maintained an extensive correspondence (over 250 letters from Buchanan are preserved in the Palace Archives). Buchanan was eventually appointed to the scientific advisory board of the Oceanographic Institute and also served as an intermediary between Albert and members of British scientific societies. See Jacqueline Carpine-Lancre and Anita McConnell, "Prince Albert and J. Y. Buchanan: Mediterranean Investigations," *History of Oceanography* 22 (January 2011): 24–31. See also Jacqueline Carpine-Lancre, "'. . . Une trace de mes anciens sillage.' Albert Ier de Monaco et la mer," in *Monaco port des Princes*, ed. André Z. Labarrère (Monaco: Yacht Club de Monaco, 1996), 44–71.

96. Prince Albert's fourth and final research vessel was *Hirondelle II*, built in France near Toulon and launched on 6 February 1911. For a description of the technological improvements made to Prince Albert's successive research vessels, as well for as an account of his work in meteorology, see Penelope Hardy, "Where Science Meets the Sea: Research Vessels and the Construction of Knowledge in the Nineteenth and Twentieth Centuries" (PhD diss., Johns Hopkins University, 2017), 111–157.
97. These northern voyages seem to have been partly inspired by Nansen's expedition on the *Fram*, and Albert consulted with Nansen in preparation for his own Arctic research. Albert I, "Croisière dans les regions arctiques," *La Grande Revue* 14 (1900): 1. See also Jacqueline Carpine-Lancre and William Barr, "The Arctic Cruises of Prince Albert I of Monaco," *Polar Record* 44, no. 1 (January 2008): 1–14.
98. Buchanan to Albert, 18 August 1896, Palace Archives, Monaco. Buchanan explains that it was he who supplied the original account to the *Times*, translated from a letter from Prince Albert.
99. "Le Prince Albert et Jules de Guerne," *Les Amis du Musée Océanographique de Monaco*, no. 32 (October 1954): 5. Richard became the first director of the oceanographic museum. He was introduced to Albert by another frequent collaborator, Jules de Guerne.
100. Albert, "Discours de S.A.S. le Prince de Monaco," 7–8.
101. As Christian Carpine has argued, dramatic improvements in oceanographic instrumentation in the late nineteenth century mirrored technological improvements in other important areas of marine knowledge: metallurgy, naval construction, and engineering. These innovations were possible in large part because of advancements in the use of steel, copper, aluminum, and plastic. Carpine, *La pratique de l'océanographie au temps du Prince Albert Ier*, 3.
102. See Christian Carpine, "Les navires océanographiques dont les noms ont été choisis par S.A.S. le Prince Albert 1er pour figurer sur la façade du Musée océanographique de Monaco," *Bulletin de l'Institut Océanographique de Monaco*, no. 2 (1968): 627–638. The care with which research vessels were selected for the facade reveals the value Albert placed on expeditionary science and the associations made between vessels and the data they produced.
103. "Généralités—L'inauguration du Musée Océanographique de Monaco," *Annales de Géographie* 19, no. 106 (1910): 374–375. The museum was also equipped with meteorological instruments. Richard to Albert, 9 October 1901, Palace Archives, Monaco. Prince Albert was also interested in meteorology and aviation. These dual interests are reflected in the Oceanographic Institute's Latin motto *Ex abyssis ad alta*. For Albert's work in meteorology, see Hardy, "Where Science Meets the Sea," 137–143.
104. See Eric Mills and Jacqueline Carpine-Lancre, "The Oceanographic Museum of Monaco," in *Ocean Frontiers, Explorations by Oceanographers on Five Continents*, ed. E. M. Borgese (New York: H. N. Abrams, 1992), 120–135.
105. "The Oceanographical Museum at Monaco," *Nature* 83, no. 2111 (14 April 1910): 192.

106. Jean Rémy Bezias, "La principauté de Monaco, la Méditerranée et la paix sous le règne du prince Albert Ier (1889–1922)," *Cahiers de la Méditerranée* 91 (2015): 1.
107. According to Jean Rémy Bezias, the institute was never particularly influential, even after its relocation to Paris in 1912. Bezias, 2.
108. Jacqueline Carpine-Lancre cites disaccord between Paul Regnard and Louis Joubin over assignment of the post of secretary general of the congress. Jacqueline Carpine-Lancre, "The Plan for an International Oceanographic Congress," in *Oceanography: The Past*, ed. M. Sears and D. Merriman (New York: Springer-Verlag, 1980), 163.
109. Carpine-Lancre, 161–162.
110. The museum's first stone was laid in 1899, and it was inaugurated in 1910.
111. For its description as a school, see "Généralités," 375. The British oceanographer William Herdman wrote of the relationship between the museum and the institute in 1923, "The factory is at Monaco, the sale-room at Paris." Herdman, *Founders of Oceanography and Their Work*, 132.
112. "Scientific Notes and News," *Science* 33, no. 841 (10 February 1911): 214.
113. For the correspondence between Albert, Murray, and Buchanan, see Jacqueline Carpine-Lancre, "Oceanographic Sovereigns: Prince Albert I of Monaco and King Carlos I of Portugal," in *Understanding the Oceans: A Century of Ocean Exploration*, ed. Margaret Deacon, Tony Rice, and Colin Summerhayes (London: CRC Press, 2001), 65.
114. "Wonders of Deep Shown in Paris," *San Francisco Chronicle*, 16 March 1913.
115. This letter is reproduced at the head of the institute's statutes. I have used the translation in Adolphe Smith, *Monaco and Monte Carlo* (London: Grant Richards, 1912), 163–164. The idea of an oceanographic institute was not entirely original. As Prince Albert was well aware, the Institut für Meereskunde opened in Berlin in 1900; Albert was present for the inauguration, along with Kaiser Wilhelm II. The Institut für Meereskunde, however—the project of German explorer Ferdinand von Richthofen (1833–1905), the Imperial German Navy, and the University of Berlin—was primarily a showcase for Germany's navy. Walter Lenz, "The Museum für Meereskunde in Berlin," *History of Oceanography: Newsletter of the ICHO* 18 (September 2006): 18. For a discussion of the Institut für Meereskunde, see Mills, *Fluid Envelope of Our Planet*, 139.
116. "D'aprés les principes en usage au Musée de Monaco." Unknown author to Prince Albert, 26 September 1910, photocopy in the Roscoff Station Archives, Roscoff, France. The letter bears the stamp of the Oceanographic Museum of Monaco Archives where, presumably, the original resides.
117. Their names and the amount they gave are still mounted on marble plaques in the aquarium room of the station. Kofoid notes that the annual rent for a table at the station at this time was 1500 francs. Kofoid, *Biological Stations of Europe*, 96.
118. When the oceanographer Marie Tharp drew her famous bathymetric charts of the world's oceans in the 1950s, she relied in part on information gathered for the General Bathymetric Chart of the World. Marie Tharp, "Connect the Dots: Mapping the Seafloor and Discovering the Mid-ocean Ridge," in *Lamont-Doherty*

Earth Observatory of Columbia: Twelve Perspectives on the First Fifty Years 1949–1999, ed. Laurence Lippsett (New York: Lamont Doherty Earth Observatory of Columbia University, 1999).

119. By 1895 two general bathymetric charts had been produced: an 1893 chart by the German navy and an 1895 chart by John Murray using *Challenger* sounding data. Jules Richard, *L'Océanographie* (Paris: Vuibert & Nony, 1907), 44.
120. Carpine-Lancre suggests that Sauerwein was woefully unprepared for this task. See Jacqueline Carpine-Lancre, "The Origin and Early History of 'La Carte Générale Bathymétrique des Océans,'" in *The History of GEBCO 1903–2003* (Utrecht, Netherlands: Lemmer, 2003), 23.
121. De Margerie also addressed a private letter to Thoulet and Sauerwein in which he questioned the "intelligence of the draughtsmen." See the excerpt reproduced in Carpine-Lancre, "The Origin and Early History of 'La Carte Générale Bathymétrique des Océans,'" 31.
122. Mills, *Fluid Envelope of Our Planet,* 175. As Mills explains, proofreading by a competent geographer might have avoided the errors, but Thoulet, the project member with the most experience, did not participate in the final production of the chart. Sauerwein seems to have been the one primarily at fault. Mills argues that Thoulet's fall from grace undermined the adoption of a mathematical approach to physical oceanography in France.
123. Improvements were made largely thanks to Henry Bourée and Alphonse Tollemer.
124. Albert 1st, *La Guerre Allemande et la Conscience Universelle* (Paris: Payot & Cie, 1919). Wilhelm II and Prince Albert had previously enjoyed a cordial personal relationship. The kaiser did not personally take up oceanographic work, but he demonstrated interest in promoting the new science in his country and visited with Albert aboard his yacht in 1898. For a description of this encounter see Raymond Damien, *Albert Ier: Prince Souverain de Monaco* (Paris: Institute de Valois, 1964), 343–346.
125. "Monaco's Ruler Applauds," *New York Times*, 27 April 1916.
126. Among the ill-fated projects from this period was a plan in 1910 to establish an international commission for the scientific exploration of the Atlantic—an organization like ICES for the Atlantic Ocean. Jens Smed, "The Decline of ICES during the First World War and Its Rise after the War," in Groeben, *Places, People, Tools,* 206.
127. He made three trips to the United States in his lifetime.
128. "Veteran Prince Thinks It's Fun to Discuss Science with American Savants," *Times Herald*, Olean, New York, 21 April 1921, 33.
129. Albert I, "Studies of the Ocean," *Scientific Monthly* 13, no. 2 (August 1921): 181.
130. Albert I, "Discours sur l'Océan," *Bulletin de l'Institut Océanographique*, no. 392 (25 June 1921): 2.
131. "The Scientist Who Ruled Monte Carlo," *Literary Digest* 74, no. 3 (15 July 1922): 38.
132. Carpine-Lancre, "The Origin and Early History of 'La Carte Générale Bathymétrique des Océans,'" 42.

133. *Hirondelle II*, one of the most advanced research vessels of its time, was put up for sale. The administrators of ICES explored the possibility of purchasing the vessel and outfitting it for a worldwide oceanographic expedition, but this project failed because of lack of funding. The *Hirondelle II* was eventually sold to an American film company. Jens Smed, "Abortive Plans for a World-Wide Oceanographic Expedition," *History of Oceanography* 12 (September 2000): 10–14. Albert wrote to Henry Maurice (president of ICES) in 1922 expressing his wish that ICES become a leader in oceanographic exploration. Otto Pettersson, "Förslag till en världsomfattande internationell havsforskningsexpedition," *Ymer* 1 (1924): 39. According to Carpine-Lancre, the vessel was eventually sold to a company in the Philippines and demolished in 1965. Jacqueline Carpine-Lancre, private communication.

3. Scientific Internationalism in a Pacific World

Epigraphs: "National Isolation Policy Scored by Secretary Work in Address at Conference," *Honolulu Advertiser,* 12 April 1927, 9; Pamela M. Henson, "The Smithsonian Goes to War: The Increase and Diffusion of Scientific Knowledge in the Pacific," in *Science and the Pacific War: Science and Survival in the Pacific, 1939–1945,* ed. Roy MacLeod (Dordrecht, Netherlands: Kluwer Academic, 2000), 44.

1. The observer was the ship's writer, Henry Cummings. Jason Smith, *To Master the Boundless Sea: The U.S. Navy, the Marine Environment, and the Cartography of Empire* (Chapel Hill: University of North Carolina Press, 2018), 108–109.
2. C. Wyville Thomson, *The Depths of the Sea: An Account of the General Results of the Dredging Cruises of the H.M.Ss. 'Porcupine' and 'Lightning'* [. . .] (London: Macmillan, 1873), 226–227.
3. Agassiz to Thomson, 6 February 1881, correspondence of Alexander E. R. Agassiz (1835–1910), University of Edinburgh Library Special Collections and Archives. Access to the Pacific Ocean was limited, not access to the Pacific coast. Alexander Agassiz traveled extensively on the west coast of South America in 1875, during which time he wrote to the members of the *Challenger* expedition then working in the Pacific. See John Murray, "Alexander Agassiz: His Life and Scientific Work," *Bulletin of the Museum of Comparative Zoology* 54, no. 3 (1911), 137–158.
4. See Larry T. Spencer, "Four Men and an Albatross: The Growth of American Oceanography, 1882–1921," in *Oceanographic History: The Pacific and Beyond,* ed. Keith Benson and Philip Rehbock (Seattle: University of Washington Press, 2002), 288–297.
5. The British had the advantage of controlling a global network of coaling depots. See Steven Gray, "Fueling Mobility: Coal and Britain's Naval Power, c. 1870–1914," *Journal of Historical Geography* 58 (2017): 92–103.
6. Sarah J. Moore, *Empire on Display: San Francisco's Panama-Pacific International Exposition of 1915* (Norman: University of Oklahoma Press, 2013), 68.
7. William A. Herdman, *Founders of Oceanography and Their Work: An Introduction to the Science of the Sea* (London: Edward Arnold, 1923), 88.

8. Herdman, 97.
9. F. Spiess, *The Meteor Expedition: Scientific Results of the German Atlantic Expedition, 1925–1927*, trans. William J. Emery (New York: Amerind, 1985), 10.
10. Spiess, 10–11. For a lengthier discussion of the *Meteor* expedition, see Penelope Hardy, "Where Science Meets the Sea: Research Vessels and the Construction of Knowledge in the Nineteenth and Twentieth Centuries" (PhD diss., Johns Hopkins University, 2017), 158–204.
11. According to Eric Mills, a lengthier voyage to the Pacific would have required refitting the coal-burning *Meteor* with diesel engines. Eric Mills, *The Fluid Envelope of Our Planet: How the Study of the Ocean Currents Became a Science* (Toronto: University of Toronto Press, 2009), 155.
12. Mills, 15, 17.
13. The first voyage of the US Fish Commission research vessel *Albatross* into the Pacific, led by Alexander Agassiz, took place in 1891. On its first voyage into the Pacific, the *Albatross* trawled near the Galápagos Islands and in the Gulf of California. Agassiz used the *Albatross* for extensive Pacific work in 1899 and 1900 on a track that took him from San Francisco, through Polynesia and Micronesia, and ended in Yokohama. Spencer, "Four Men and an Albatross," 290–291. See also Walter Lenz, "The Aspirations of Alfred Merz, George Wüst, and Albert Defant: From Berlin to Pacific Oceanography," in Benson and Rehbock, *Oceanographic History*, 118–123.
14. Theodore M. Knappen, "The Sinews of War: California Fish Industry Has Been Stimulated to Undreamed-of Proportions," *New York Tribune*, 10 August 1917.
15. A. P. Elkin, "Pacific Science Association: Its History and Role in International Cooperation," in *Bernice P. Bishop Museum Special Publication*, no. 48 (Hawaii: Bishop Museum Press, 1961), 7.
16. Roy Elwood Clausen, "The Sixth Pacific Science Congress," *Science* 90, no. 2342 (17 November 1939): 449–456.
17. Herbert H. Gregory, "Address," in *Fifth Pacific Science Congress Proceedings* (Vancouver: Pacific Science Association, 1933), 1:131.
18. "Public Sessions," *Sixth Pacific Science Congress Proceedings* (Berkeley: University of California Press, 1940), 1:49.
19. "Public Sessions," 51.
20. Historians of ocean science have, for instance, emphasized the military origins of post–World War II Pacific-based scientific programs like those at the Scripps Institution of Oceanography. See Jacob Darwin Hamblin, *Oceanographers and the Cold War: Disciples of Marine Science* (Seattle: University of Washington Press, 2005).
21. A 1924 issue of the *Pan-Pacific Union Bulletin* describes the proposed function of the Pan-Pacific Science Institute as "a clearing house for scientific information." G. E. Allen, "Pan-Pacific Scientific Research Work," *Pan-Pacific Union Bulletin*, no. 61 (November 1924): 3–4.
22. Helen Rozwadowski, "Internationalism, Environmental Necessity, and National Interest: Marine Science and Other Sciences," *Minerva* 42 (2004): 127–149.
23. Jacob Hamblin has made a similar claim in his study of Cold War oceanography; he writes that "the lack of national borders at sea, the indiscriminately hostile

environmental conditions, and the global scope of observations have long lent oceanography the reputation of being an inherently international endeavor." Hamblin, *Oceanographers and the Cold War*, xix.
24. Frank Bergon, ed., *The Journals of Lewis and Clark* (New York: Penguin Books, 2003), xxiv.
25. Charles Wilkes, *Voyage Round the World* [...] (Philadelphia: George W. Gorton, 1849), viii; Charles Wilkes, *Synopsis of the Cruise of the U.S. Exploring Expedition* [...] (Washington, DC: Peter Force, 1842), 52.
26. H. B–, "Scraps from the Lucky Bag," *Southern Literary Messenger* (April 1840): 238.
27. "Japan Opened: Satisfactory Result of Commodore Perry's Visit...," *New York Times*, 13 June 1854.
28. "Resolutions Adopted at the Fourth Pacific Science Congress," in *Fifth Pacific Science Congress Proceedings* 1:250. Twenty-five countries were represented at the Fourth Pacific Congress meeting.
29. "Resolutions Adopted at the Fourth Pacific Science Congress," 1:254.
30. "Resolutions Adopted at the Fourth Pacific Science Congress," 1:249. Vaughan was himself instrumental in transforming the Scripps Institution into a center for oceanographic, and not solely biological, research after he became director in 1924. See Ki Won Han, "The Rise of Oceanography in the United States, 1900–1940" (PhD diss., University of California, Berkeley, 2010), 92–97.
31. See Thomas Wayland Vaughan, *International Aspects of Oceanography: Oceanographic Data and Provisions for Oceanographic Research* (Washington, DC: National Academy of Sciences, 1937). This report was prepared at the behest of the National Academy of Sciences. Vaughan began compiling the information for his report in 1934, but publication was delayed by illness until 1937.
32. Scripps Archives, George McEwen Papers and Correspondence (MC21), box 7, folder 6.
33. Eric Mills, "Useful in Many Capacities. An Early Career in American Physical Oceanography," *Historical Studies in the Physical and Biological Sciences* 20, no. 2 (1990): 268.
34. George McEwen, "The Distribution of Temperature and Salinities, and the Circulation in the North Pacific Ocean," *Bulletin of the Scripps Institution for Biological Research of the University of California*, no. 9 (15 December 1919): 63.
35. An international fisheries commission, establishing a partnership between the United States and Canada, was founded in 1923. It was largely a regional organization and was later renamed the International Pacific Halibut Commission. A true transpacific fisheries research program did not take shape until the years after World War II, with the formation of the International North Pacific Fisheries Commission by Canada, the United States, and Japan in 1951. Sara Tjossem, *The Journey to PICES: Scientific Cooperation in the North Pacific* (Fairbanks: University of Alaska, Alaska Sea Grant Program, 2005), 16–17. See also Lissa Wadewitz, *The Nature of Borders: Salmon, Boundaries, and Bandits on the Salish Sea* (Seattle: University of Washington Press, 2012), 158–167; Carmel Finley, *All the Fish in the Sea: Maximum Sustainable Yield and the Failure of Fisheries Management* (Chicago: University of Chicago Press, 2011), 134–135.

36. "To Make Tour of World," *San Francisco Chronicle*, January 28, 1906.
37. See Antony Adler, "A Letter from Japan: Internationalism and Pacific Oceanography," *Bulletin of the Pacific Circle*, April 2014, 32. For international cooperation and Pacific oceanography during this period, see Eric Mills, "The Oceanography of the Pacific: George F. McEwen, H. U. Sverdrup and the Origin of Physical Oceanography on the West Coast of North America," *Annals of Science* 48 (1991): 241–266. See also Philip Rehbock, "Organizing Pacific Science: Local and International Origins of the Pacific Science Association," *Pacific Science* 45, no. 2 (1991): 107–122. Russian scientists also increased survey work in the eastern Pacific, spearheaded by the Pacific Research Institute of Fisheries and Oceanography, established in Vladivostok in 1929. N. N. Mikhailov, E. D. Vyazilov, V. I. Lomonov, N. S. Studyonov, and M. Z. Shaimardanov, "Russian Marine Expeditional Investigations of the World Ocean," *International Ocean Atlas and Information Series* (Silver Spring, MD: National Oceanic and Atmospheric Administration, 2002), 5:37.
38. Charles McLean Fraser, "Oceanography: Marine Zoology," *Scientific Monthly* 44, no. 1 (January 1937): 62. Born in Ontario in 1872, Charles McLean Fraser served as curator of the Pacific Biological Station at Nanaimo from 1912 to 1924. He became a professor at the University of British Columbia in 1920. For more on Fraser, see Kenneth Johnstone, *The Aquatic Explorers: A History of the Fisheries Research Board of Canada* (Toronto: University of Toronto Press, 1977), 92.
39. Fraser, "Oceanography," 62.
40. Eckhardt Fuchs writes that Pacific "internationalism . . . found its institutional expression in the Pan-Pacific Union, founded by Alexander H. Ford in 1917. . . . Under the doctrine of 'Patriotism of the Pacific,' the primary goal of the Pan-Pacific Union was to promote mutual cultural understanding among Pacific peoples. . . . Ford's idea of holding international congresses to create this 'international brotherhood' coincided with [William Alanson] Bryan's efforts toward scientific exploration of the Pacific." Eckhardt Fuchs, "The Politics of the Republic of Learning: Internationalism Scientific Congresses in Europe, the Pacific Rim, and Latin America," in *Across Cultural Borders: Historiography in Global Perspective*, ed. Eckhardt Fuchs and Benedikt Stuchtey (Lanham, MD: Rowman & Littlefield, 2002), 217–218. See also Roy MacLeod and Philip Rehbock, "Developing a Sense of the Pacific: The 1923 Pan-Pacific Science Congress in Australia," *Pacific Science* 54, no. 3 (2000), 209–225.
41. William Alanson Bryan, "The Pacific Scientific Institution: An Address by William Alanson Bryan," in *Pacific Institution Publications*, special series no. 2 (Chicago: 1908), 5.
42. Bryan, 8.
43. This interpretation of nature in the Pacific was not entirely new. Johann Reinhold Forster, the naturalist who accompanied Cook on his second Pacific voyage, declared in 1778 that his object was "an investigation of Nature, in its greatest extent; the earth, the sea, the air, the organic and animated creation, and more particularly mankind itself, so far as all these came within the reach of examination, in the course of a voyage round the world." Johann Reinhold Forster,

Observations Made during a Voyage Round the World, ed. Nicholas Thomas, Harriet Guest, and Michael Dettelbach (Honolulu: University of Hawaii Press, 1996), lxxvii.

44. As Roy MacLeod and Philip Rehbock have argued, "In the Pacific lay opportunities for phenomena to be observed and theories to be tested. What applied to botany and geology applied equally to native peoples—as it would to oceanography, meteorology, marine biology, and eventually nuclear weapons." Roy MacLeod and Philip Rehbock, preface to *Darwin's Laboratory: Evolutionary Theory and Natural History in the Pacific* (Honolulu: University of Hawaii Press, 1994), ix.
45. R. A. Daly, "Problems of the Pacific Islands," *American Journal of Science* 41 (February 1916): 154.
46. Daly, 154.
47. H. V. Neal, "Report of the San Francisco Meeting of Section F of the American Association for the Advancement of Science," *Science* 42, no. 1087 (29 October 1915): 619.
48. "Where Dull Care Is an Unknown Quantity," *Los Angeles Times*, 4 April 1915.
49. Robert Rydell, *World of Fairs: The Century-of-Progress Expositions* (Chicago: University of Chicago Press, 1993), 85.
50. Carolyn Anspacher, "Pacific House Popular," *San Francisco Chronicle, Exposition Extra*, 20 February 1939.
51. Philip N. Youtz, "Proposal for a Permanent Pacific House," manuscript, San Francisco Public Library Special Collections, GGIE folder, 2.
52. Youtz, 2.
53. Franklin D. Roosevelt, "Fireside Chat on Progress of the War," *Public Papers of the Presidents of the United States: F. D. Roosevelt, 1942* (New York: Harper, 1950), 11:106. FDR, like his uncle Theodore, served as secretary of the navy prior to his presidency. As secretary, FDR was a vocal advocate for the expansion of American naval power in the Pacific. See Frank Freidel, *Franklin D. Roosevelt: The Apprenticeship* (Boston: Little, Brown, 1952), 220–235.
54. Jack James and Earle Weller, *Treasure Island, "The Magic City," 1939–1940: The Story of the Golden Gate International Exposition* (San Francisco: Pisani Printing, 1941), 17.
55. R. W. Cary, *The Naval History of Treasure Island* (Treasure Island, CA: US Naval Training and Distribution Center, 1946), 25; ellipsis in original. There is some evidence to suggest that similar reframing of the fair occurred in Japan. Katherine Caldwell, who served as the director of education at the Palace of Fine Arts during the fair, later recalled, "The terrible thing was that some of the people in Japan who lent their treasures were accused of being pro-American. One of them, an elderly collector of refinement, who was interested in art, not politics, was jailed for a while." "Katherine Caldwell," *The San Francisco Fair: Treasure Island 1939–1940*, ed. Patricia F. Carpenter and Pail Totah (San Francisco: Scottwall Associates, 1989), 47.
56. Carpenter and Totah, *The San Francisco Fair*, 147.
57. Carpenter and Totah, 56.

58. Stephen Schlesinger, *Act of Creation: The Founding of the United Nations* (Boulder, CO: Westview, 2003), 61.
59. Schlesinger, 117.
60. "Address of Welcome Given by the President, Dr. R. A. Falla," *Seventh Pacific Science Congress Proceedings* (Wellington: R. E. Owen Government printer, 1951), 1:46.
61. As one example, the Woods Hole plankton biologist Mary Sears, given the task of providing oceanographic science intelligence in advance of troop landings on Pacific islands, relied heavily on captured Japanese hydrographic charts. Much has been written about the wartime work of Mary Sears, but see, in particular, Kathleen Broome Williams, "From Civilian Planktonologist to Navy Oceanographer: Mary Sears in World War II," in *The Machine in Neptune's Garden: Historical Perspectives on Technology and the Marine Environment*, ed. Helen Rozwadowski and David K. van Keuren (Sagamore Beach, MA: Science History, 2004), 243–272. See also Susan Schlee, *The Edge of an Unfamiliar World: A History of Oceanography* (New York: E. P. Dutton, 1973), 311.
62. Katsuma Dan, who had worked at the Marine Biological Laboratory in Woods Hole and was married to the American biologist Jean M. Clark, was best known in scientific circles for his work on cell biology.
63. John Findlay, *Magic Lands: Western Cityscapes and American Culture after 1940* (Berkeley: University of California Press, 1992), 214.
64. Vannevar Bush was head of the US Office of Scientific Research and Development during World War II and was instrumental in the creation of the National Science Foundation.
65. Vannevar Bush, *Science, the Endless Frontier: A Report to the President* (Washington, DC: National Science Foundation, 1990), 5.
66. Findlay, *Magic Lands*, 228.
67. As the historian James Spiller has argued, "America and the Soviet Union each vied for international leadership by offering the fruits of its hard power and the allure of its utopian ideology." At the same time, "the Cold War was a hegemonic project for the United States," which "could not sustainably exert international leadership without the abiding support of people at home and abroad." James Spiller, *Frontiers for the American Century: Outer Space, Antarctica, and Cold War Nationalism* (New York: Palgrave Macmillan, 2015), 7.
68. NASA, *Astronautical and Aeronautical Events of 1962: Report of the National Aeronautics and Space Administration to the Committee on Science and Astronautics U.S. House of Representatives Eighty-Eight Congress, First Session* (Washington, DC: Government Printing Office, June 1963), 74.
69. NASA, 87.
70. Yamasaki, whose family had been interned during the war, had previously gained national recognition as the architect of the US exhibit at the 1959 World Agricultural Fair in New Delhi, India.
71. He had completed the doctorate degree work in 1935, but the award of the degree seems to have been postponed by the war. For the chronology of Spilhaus's career, I am relying on a short biography provided by Woods Hole. "MC-46:

Athelstan Frederick Spilhaus papers," Woods Hole Oceanographic Institution Data Library and Archives, http://dla.whoi.edu/manuscripts/node/196143.
72. William A. Nierenberg, "Athelstan Spilhaus, 25 November 1911–30 March 1998," *Proceedings of the American Philosophical Society* 144, no. 3 (September 2000): 344.
73. Athelstan Spilhaus, "Oceanography: A Wet and Wondrous Journey," *Bulletin of the Atomic Scientists: Journal of Science and Public Affairs*, December 1964, 12–13. As discussed further in Chapter 4, Spilhaus was not alone in holding such utopian underwater dreams. Statements like these likely inspired Arthur C. Clark's 1957 novel *The Deep Range*.
74. Sharon Moen, *With Tomorrow in Mind: How Athelstan Spilhaus Turned America toward the Future* (Minneapolis: University of Minnesota, 2015), 67–80.
75. Warren G. Magnuson, "Magnuson Preview: Century 21 Science Pavilion to Be 'Jewel Box' of Wonder," *Seattle Post-Intelligencer*, 5 March 1961. Not all shared this optimistic view of science. See, for instance, the science fiction writer Robert Heinlein's critique quoted in Erik Ellis, "Dixy Lee Ray, Marine Biology, and Public Understanding of Science in the United States (1930–1970)" (PhD diss., Oregon State University, 2005), 180.
76. Athelstan Spilhaus, *U.S. Science Exhibit Seattle World's Fair: Final Report* (Washington, DC: US Department of Commerce, 1963), 7.
77. "The House of Science," US National Archives. https://www.youtube.com/watch?v=ivuOyD1BIH0.
78. Spilhaus, 48.
79. Spilhaus, 41.
80. Findlay, *Magic Lands*, 238.
81. Carolyn Bennett Patterson, "Seattle Fair Looks to the 21st Century," *National Geographic* 122, no. 3 (September 1962): 420–424.
82. Ellis, "Dixy Lee Ray," 186–187.
83. The science exhibits did not close with the termination of the fair in 1962, an NSF grant of $100,000 allowing some exhibits to remain open. Ellis, 187.
84. Ellis, 188.
85. After her brief tenure as director of the Science Center, Ray went on to serve as a chief scientist with the International Indian Ocean Expedition (1962–1965). Rachel White Scheuering, *Shapers of the Great Debate: A Biographical Dictionary* (Westport, CT: Greenwood, 2004), 92. See also Hamblin, *Oceanographers and the Cold War*, 120–127. For Dixy Lee Ray's critique of the environmental movement, see Naomi Oreskes and Erik M. Conway, *Merchants of Doubt: How a Handful of Scientists Obscured the Truth on Issues from Tobacco Smoke to Global Warming* (New York: Bloomsbury, 2010), 130.
86. Initial efforts to forge collaboration between oceanographers and the US Navy predate World War II. See Tjossem, *Journey to PICES*, 4.
87. See, for example, Martha Smith-Norris, "'Only as Dust in the Face of the Wind': An Analysis of the BRAVO Nuclear Incident in the Pacific, 1954," *Journal of American–East Asian Relations* 6, no. 1 (Spring 1997): 1–34.
88. Elizabeth Noble Shor, *Scripps Institution of Oceanography: Probing the Oceans 1936 to 1976* (San Diego, CA: Tofua Press, 1978), 36.

89. Ronald Rainger, "'A Wonderful Oceanographic Tool': The Atomic Bomb, Radioactivity and the Development of American Oceanography," in Rozwadowski and van Keuren, *Machine in Neptune's Garden*, 93–131.
90. Walter Munk, "Glimpses of Oceanography in the Postwar Period," *Oceanography* 21, no. 3 (September 2008): 19.
91. See, for example, John Steinbeck's plea for increased funding for oceans exploration. Steinbeck suggested that the ocean exploration technology offered a more peaceful outlet for Cold War innovation than rockets. John Steinbeck, "Let's Go After the Neglected Treasures beneath the Seas: A Plea for Equal Effort on 'Inner Space' Exploration," *Popular Science*, September 1966, 84–87.
92. Hamblin, *Oceanographers and the Cold War*, xviii.
93. Hamblin, 260.
94. See Helen Rozwadowski, "Arthur C. Clarke and the Limitations of the Ocean as a Frontier," *Environmental History* 17, no. 3 (2012): 578–602.
95. John F. Kennedy, State of the Union Address, 30 January 1961.
96. "Ocean Frontier," *Time*, 6 July 1959, 44.
97. "Ocean Frontier," 44. For Iselin and his prewar role as intermediary between oceanographers and the navy, see Gary E. Weir, "Fashioning Naval Oceanography: Columbus O'Donnell Iselin and American Preparation for War, 1940–1941," in Rozwadowski and van Keuren, *Machine in Neptune's Garden*, 65–91.
98. Shor, *Scripps Institution of Oceanography*, 85–86. For an example of large-scale naval acoustic surveillance, see Gary Weir, "The American Sound Surveillance System: Using the Ocean to Hunt Soviet Submarines, 1950–1961," *International Journal of Naval History* 5, no. 2 (August 2006). http://www.ijnhonline.org/wp-content/uploads/2012/01/article_weir_aug06.pdf.
99. Shor, 34.
100. See, for example, Naomi Oreskes, "A Context of Motivation: US Navy Oceanographic Research and the Discovery of Hydrothermal Vents," *Social Studies of Science* 33, no. 5 (October 2003): 697–742.
101. The international collaborative oceanographic surveys of the postwar period include the International Geophysical Year of 1956–1959, the International Indian Ocean Expedition (1962–1967), and the International Decade of Ocean Exploration (1971–1980).
102. Edward Wenk, *The Politics of the Ocean* (Seattle: University of Washington Press, 1972), 212.
103. A good example is Roger Revelle's efforts to solve the problem of desalinating agricultural irrigation water in Pakistan in the early 1960s—described by him as an alternative to giving weapons to the Pakistani military. See "Interview with Dr. Roger Revelle, La Jolla, CA, 3 February 1989," *American Institute of Physics*, http://www.aip.org/history/ohilist/5051.html.

4. Cold War Science on the Seafloor

Epigraphs: Edward Wenk Jr., *The Politics of the Ocean* (Seattle: University of Washington Press, 1972), 52; Helen Rozwadowski, "From Danger Zone to World of

Wonder: The 1950s Transformation of the Ocean's Depths," *Coriolis* 4, no. 1 (2013): 16.

1. Here I focus on a clearly delineated marine space, but my broader aim is to provide an interpretive approach that can be applied to any marine space where human activity takes place.
2. "Message to Congress: Johnson's Conservation Message," in *CQ Almanac 1968*, 24th ed., 20-61-A-20-67-A (Washington, DC: Congressional Quarterly, 1969), http://library.cqpress.com/cqalmanac/cqal68-1284522.
3. For example, see Joseph Hromadik and Robert Breckenridge, "Construction Concepts for the Deep Ocean," in *Proceedings of the Conference on Civil Engineering in the Oceans, ASCE Conference, San Francisco, California, September 6–8, 1967* (New York: American Society of Civil Engineers, 1968): 713–739.
4. Yet as Helen Rozwadowski points out, "Decline or abandonment of technologies and technological systems, as well as false starts, can offer insights that would be lost by studying only the establishment of those technologies that we currently use." Helen Rozwadowski, "Engineering, Imagination, and Industry: Scripps Island and Dreams for Ocean Science in the 1960s," in *The Machine in Neptune's Garden*, ed. Helen Rozwadowski and David K. van Keuren (Sagamore Beach, MA: Science History, 2004), 317.
5. Wartime oceanographic work sponsored by the navy at the Woods Hole Oceanographic Institution developed military applications for the bathythermograph invented by Athelstan Spilhaus. The navy used Spilhaus's bathythermograph to calibrate submarine-detecting sonar equipment, compensating for detection interference caused by variations in temperature and pressure. As Todd A. Wildermuth has put it, the bathythermograph "rendered water more legible." Todd Wildermuth, "Yesterday's City of Tomorrow: The Minnesota Experimental City and Green Urbanism" (PhD diss., Dept. of Natural Resources and Environmental Sciences, University of Illinois, Urbana-Champaign, 2008), 25. For his part, Spilhaus described the bathythermograph as "a product of laziness" because it allowed continuous measurement while a vessel was underway. Victor Cohn, "Spilhaus 'Lazed' His Way to Oceanography Landmark," *Minneapolis Sunday Tribune*, 28 June 1959, 5.
6. The first undergraduate oceanography program in the United States began at the University of Washington in 1928. Eight universities included oceanography in their curriculum by 1946; the number of universities offering oceanography programs doubled between 1957 and 1963. Roger H. Charlier, "Growth of Oceanographic Education in the United States," *Limnology and Oceanography* 11, no. 4 (April 1966): 637.
7. Gary Weir likens these individuals to "translators," capable of "cross-cultural communication." Gary Weir, *An Ocean in Common: American Naval Officers, Scientists, and the Ocean Environment* (College Station: Texas A&M University Press, 2001), xi.
8. As Ronald Rainger argues, the foremost American marine scientists of the 1920s, Thomas Wayland Vaughan and Henry Bryant Bigelow, considered oceanography to be "the study of a particular place—the oceans; and oceanographers were

those who used the tools of geology, biology, physics, and chemistry to examine the seas." Ronald Rainger, "Constructing a Landscape for Postwar Science: Roger Revelle, the Scripps Institution and the University of California, San Diego," *Minerva* 39, no. 3 (2001): 329.

9. As Naomi Oreskes writes, "If the issue of the effect of science on the Cold War has long been argued, the reverse question (How did the Cold War affect science?) did not receive sustained academic scrutiny until relatively recently." Naomi Oreskes, "Science in the Origins of the Cold War," in *Science and Technology in the Global Cold War*, ed. Naomi Oreskes and John Krige (Cambridge, MA: MIT Press, 2014), 16. Significantly, as Rozwadowski has argued, the 1950s may be considered a "moment of sudden dramatic cultural change in human perception of the undersea environment." Helen Rozwadowski, "From Danger Zone to World of Wonder: The 1950s Transformation of the Ocean's Depths," *Coriolis*, 4, no. 1 (2013): 1.

10. Marita Sturken and Douglas Thomas, introduction to *Technological Visions: The Hopes and Fears That Shape New Technologies*, ed. Marita Sturken, Douglas Thomas, and Sandra J. Ball-Rokeach (Philadelphia: Temple University Press, 2004), 1.

11. John E. Crayford, *Underwater Work: A Manual of Scuba Commercial, Salvage and Construction Operations* (Cambridge, MD: Cornell Maritime Press, 1959), 172.

12. Crayford, 120.

13. Crayford, 183.

14. See Rozwadowski, "From Danger Zone to World of Wonder"; and Eric Hanauer, *Diving Pioneers: An Oral History of Diving in America* (San Diego, CA: Watersport, 1994), 140.

15. A comparison can be made between the history of diving and the history of computing. As Paul Ceruzzi has argued, personal computing was perfected through the work of electronics hobbyists and enthusiasts who experimented with microprocessor-based systems in much the same way that radio amateurs experimented with high-frequency communication a generation earlier. Similarly, with the arrival of scuba to the United States, amateur diving enthusiasts immediately set about tinkering with the new technology. The invention of scuba was itself the indirect result of wartime tinkering. The breathing regulator was adapted from a car engine valve modification developed by Gagnan to allow a vehicle to burn cooking oil instead of regular gasoline—a requirement due to the wartime shortages in Vichy France. Paul E. Ceruzzi, *A History of Modern Computing*, 2nd ed. (Cambridge, MA: MIT Press, 2003), 224.

16. Wheeler J. North interviewed by Shelley Erwin, Pasadena, California, 6 October–1 December 1998. Oral History Project, California Institute of Technology Archives, http://resolver.caltech.edu/CaltechOH:OH_North_W.pdf, 41.

17. Rozwadowski, "From Danger Zone to World of Wonder," 9–14. The invention of the wet suit is credited to the physicist Hugh Bradner.

18. Rozwadowski cautions, "It is impossible to separate diving technology and consider it apart from the people who used it. Conversely, it is equally inadvisable to write the history of the ocean environment apart from the technology

that provided access to it." Rozwadowski, "From Danger Zone to World of Wonder," 15.
19. Willard N. Bascom and Roger Revelle, "Free-Diving: A New Exploratory Tool," *American Scientist* 41, no. 4 (October 1953): 624.
20. Eric Hanauer explains, "From 1952 to 1960, Scripps was the epicenter of diving for the University of California, conducting all the training for the other campuses, certifying divers, and maintaining all the records." Eric Hanauer, "Scientific Diving at Scripps," *Oceanography* 16, no. 3 (2003): 91.
21. Alister Hardy, "Will Man Be More Aquatic in the Future?," *New Scientist*, 24 March 1960, 730–733.
22. See Warren Belasco, *Meals to Come: A History of the Future of Food* (Berkeley: University of California Press, 2006), 205.
23. See, for example, P. M. Borisov, "Can We Control the Arctic Climate?," *Bulletin of the Atomic Scientists*, March 1969, 43–48. Originally published in the Soviet journal *Priroda*, Borisov's article was subsequently translated by the Canadian Defense Research Board.
24. The historian Gary Kroll has argued, "Cousteau created an ocean that was easily explored and imminently habitable through the genius of science and technology." Gary Kroll, *America's Ocean Wilderness: A Cultural History of Twentieth-Century Exploration* (Lawrence: University of Kansas Press, 2008), 7.
25. George Robeson, "Cousteau Is Not in the Hotel Business," *Independent (Long Beach, California)*, 18 April 1969.
26. The museum's third director, Cousteau seems primarily to have thought of the museum as a base of operations and a depository for collections gathered during his expeditions aboard the *Calypso*. See Jacques-Yves Cousteau, "Les adieux du Musée Océanographique à son Directeur le commandant Jules Rouch," *Les amis du Musée Océanographique de Monaco*, no. 41 (1957): 4.
27. Quoted in Brad Matsen, *Jacques Cousteau: The Sea King* (New York: Pantheon, 2009), 160–161.
28. Wallace Cloud, "The Race for the Bottom of the Sea," *Popular Science*, July 1963, 36. Not all contemporary ocean exploration enthusiasts shared Cousteau's optimism. In 1973, Hans Hass, the Austrian diving pioneer, wrote that he did not "see the slightest chance of realizing Cousteau's vision." Hans Hass, *Men beneath the Sea: Man's Conquest of the Underwater World* (New York: St. Martin's Press, 1975), 374.
29. Matsen, *Jacques Cousteau*, 161–162.
30. Jacques-Yves Cousteau, "Ocean-Bottom Homes for Skin Divers," *Popular Mechanics*, July 1963, 98.
31. "Seabed City of the Future," *Sydney Morning Herald*, 19 January 1963, 11.
32. Many of Cousteau's projects were backed by external investors. His famous research vessel *Calypso* and his film *The Silent World* were funded by British politician and magnate Loel Guinness. It was rumored that Cousteau collaborated with the Central Intelligence Agency. Axel Madsen attributes the origin of these stories to the crew of the American Red Cross ship *Hope*, which observed the *Calypso*—equipped with US Navy instruments—sailing into territorial waters

they were themselves forbidden to enter. Axel Madsen, *Cousteau: An Unauthorized Biography* (New York: Beaufort Books, 1986), 210. For more on Cousteau's film work, see Graham Huggan, *Nature's Saviours: Celebrity Conservationists in the Television Age* (London: Routledge, 2013), 65–104.
33. Quote appears in "Seabed City of the Future," *The Sydney Morning Herald*, 11. See also Cousteau's account of Conshelf I, which includes excerpts from Falco's journal, in Cousteau, "Ocean-Bottom Homes for Skin Divers." Falco later recalled the physiological effect of switching from breathing oxygen to heliox, a mixture of oxygen and helium developed specifically for saturation diving. "On air, we find everything so beautiful, but with heliox, the reality is there, gray and sad." Jacques Cousteau and Alexis Sivirine, *Jacques Cousteau's Calypso* (New York: H. N. Abrams, 1983), 82.
34. Madsen, *Cousteau*, 128. It should be noted that Cousteau held racist views, but he did not voice them publicly or act on them. His elder brother, Pierre-Antoine, worked as a propagandist for the Vichy government during the war, publishing virulently anti-Semitic texts. Although Cousteau later sought to distance himself from his brother, and played down the privileged status he himself had enjoyed under Vichy, one wonders how his imagined aquatic man of the future fit in a worldview that denied equal worth to all men.
35. "Membrane Filters Air," *Science News Letter*, 7 November 1964, 293.
36. The patents were submitted by Waldemar ("Wally") A. Ayers and Lewis H. Strauss, son of Lewis L. Strauss who was a former secretary of commerce and a member of the Atomic Energy Commission. See also Stacy V. Jones, "Physicist Develops a New Breathing Apparatus for Underwater Swimming," *New York Times*, 13 May 1967.
37. Wallace Cloud, "No Snorkel, No Scuba. Artificial Gills: They'll Let You Breathe like a Fish," *Popular Mechanics*, December 1967, 69.
38. An image of a liquid-breathing mouse from Kylstra's experiments appeared in *Life Magazine*. "A Mouse Breathes Liquid—And Lives," *Life Magazine* (25 August 1967), 77.
39. See "Man Can Function in Deep Water for Extended Period of Time," *Health Bulletin* 84, no. 3 (March 1969): 7.
40. "Letter to Edward Lanphier from Edwin Link," John H. Evans Library Digital Collections, https://digcollections.lib.fit.edu/items/show/1249.
41. Conshelf I took place in September 1962, and Conshelf II took place in June 1963. A near-fatal accident occurred during the installation of the Conshelf II Deep Cabin. Madsen, *Cousteau*, 131.
42. The Mysterious Island was designed by Cousteau and built under the direction of the famed Swiss engineer and explorer Jacques Piccard. The two worked together on several occasions, Piccard sometimes joining *Calypso* expeditions. See, for example, André Laban, Jean-Marie Pérès, and Jacques Piccard, "La photographie sous-marine profonde et son exploitation scientifique," *Bulletin de l'Institut Océanographique de Monaco* 60, no. 1258 (March 1963): 1–32.
43. Daniel Behrman, *Exploring the Ocean* (Paris: UNESCO, 1970), 24.
44. Behrman, 15.

45. Already in 1959 a writer for the Associated Press speculated that seabed metals could "break the world market." Rennie Taylor, "Ocean Bottoms Are Loaded with Loot of Rare Metals," *Santa Cruz Sentinel*, 18 February 1959, 12.
46. Part of the navy's Deep Submergence Systems Project, Sealab I launched off Bermuda in 1964, and Sealab II was deployed in the La Jolla Canyon off the coast of Southern California in 1968. Sealab III launched in February of 1969 but was scrapped after the death of an aquanaut shortly after deployment. When Sealab III closed, the navy's saturation diving program went fully undercover. "The Projects," as these top-secret missions were called, took place in the North Pacific, off the coast of Siberia, and deployed underwater communication surveillance equipment. Ben Hellwarth, *Sealab: America's Forgotten Quest to Live and Work on the Ocean Floor* (New York: Simon & Schuster, 2012).
47. See, for example, a prediction for the future domestic life of the middle-class "200-foot family" living under the sea in Coles Phinizy, "Settlers at the Bottom of the Sea," *Sports Illustrated*, 29 June 1964, 36.
48. Wallace Cloud, "Science 'Cities' under the Sea," *Popular Science*, February 1968, 123.
49. Helen Rozwadowski, *Vast Expanses: A History of the Ocean* (London: Reaktion Books, 2018), 205.
50. Nor were these visions restricted to the American imagination. To give but one example, designs for floating cities were promoted in Japan in the late 1950s and 1960s. The works of architects Masato Otaka, Kiyonori Kikitake, and Noriaki Kisho Kurokawa exemplify this movement.
51. In the Atlantic, Gregg Seamount, midway between Bermuda and the Grand Banks, was the focus of a scientific expedition in 1967 carried out by the Environmental Science Service Administration. "Submerged Mountain Will Be Scene of Big Expedition," *Clarion-Ledger* (Jackson, Mississippi), 16 July 1967.
52. Austin pointed out that coal mines in Newfoundland (Bell Island) and Nova Scotia (Cape Breton) already extended beneath the seabed, demonstrating the feasibility of "rock site" installations. C. F. Austin, "Manned Undersea Structures: The Rock-Site Concept," *U.S. Naval Ordnance Test Station Technical Paper 4162* (China Lake, California: U.S. Naval Ordnance Test Station, October 1966).
53. Nielkanth J. Shen-D'Ge, "Structures/Materials Synthesis for Safety of Oceanic Deep-Submergence Bottom-Fixed Manned Habitat," *Journal of Hydronautics* 2, no. 3 (1968): 120–130.
54. National Commission on Marine Science, Engineering, and Resources, *Our Nation and the Sea: A Plan for National Action* (Washington, DC: Government Printing Office, January 1969), 49.
55. National Commission on Marine Science, 70.
56. National Commission on Marine Science, 72.
57. National Commission on Marine Science, 161–162.
58. "Hague Deadlocked over 3-Mile Limit," *New York Times*, 8 April 1930. For President Wilson's insistence on maintaining "freedom of the seas" as part of his

fourteen-point peace treaty at the end of World War I and British resistance on this point, see Margaret MacMillan, *Paris 1919: Six Months That Changed the World* (New York: Random House, 2002), 13, 19.

59. Peter Calvert, *The International Politics of Latin America* (Manchester, UK: Manchester University Press, 1994), 203–205. For postwar fisheries competition in the Pacific, see Carmel Finley, *All the Boats on the Ocean: How Government Subsidies Led to Global Overfishing* (Chicago: University of Chicago Press, 2017).
60. Nikos Papadakis, *The International Legal Regime of Artificial Islands* (Leiden, Netherlands: Sijthoff Ocean Development, 1977), 181. This legislation, as well as the 1953 Submerged Lands Act, resulted in part from the Supreme Court's decision in favor of the federal government in the 1947 case United States vs. California. See U.S. Reports: United States v. California, 332 U.S. 19 (1947) (Hugo Lafayette Black, Judge), https://www.loc.gov/item/usrep332019/.
61. Max Frankel, "Legal Chaos Caused by Conflict on Extent of Territorial Waters," *New York Times*, 10 June 1963.
62. Pirate stations were suppressed with the passing of the European Agreement for the Prevention of Broadcasts Transmitted from Stations Outside National Territories by the European Council in 1965. The most famous of the pirate stations was probably the Nordzee television and radio broadcast station housed on an Irish-built platform off the coast of the Netherlands. Dutch marines raided the platform, rappelling down from a helicopter, and put an end to the broadcasts in December 1964.
63. "Appeals Court Sinks Builder's Hope for 'Republic' on Reef," *Miami News*, 23 January 1970.
64. "Judge Outlaws Island Nations off Florida Coast," *Florida Today*, 4 January 1969.
65. Ann L. Hollick and Robert E. Osgood, *New Era of Ocean Politics* (Baltimore: Johns Hopkins University Press, 1974), 18.
66. Evan Luard, *The Control of the Sea-Bed: A New International Issue* (London: Heinemann, 1974), 56.
67. Luard, 20.
68. Luard, 21.
69. Luard, 24.
70. Luard, 23–25.
71. See Richard B. Bilder, "The Canadian Arctic Waters Pollution Prevention Act: New Stresses on the Law of the Sea," *Michigan Law Review Association* 69, no. 1 (1970): 1–54.
72. The details of negotiations leading up to the formulation of this memorandum are provided in Hollick and Osgood, *New Era of Ocean Politics*, 29–37.
73. The proposal was a compromise between the State Department and the Defense Department, preserving local revenue but aiming to curtail jurisdictional creep and seabed militarization. More negotiations between State Department and petroleum industry representatives followed this announcement. A formal treaty, the United Nations Draft Convention on the International Seabed Area, was finally presented to the UN on August 3, 1970. Hollick and Osgood, 39.

74. Clyde Sanger, *Ordering the Oceans: The Making of the Law of the Sea* (Toronto: University of Toronto Press, 1987), 125.
75. William T. Burke, "Marine Science Research and International Law," Law of the Sea Institute, occasional paper no. 8 (September 1970): 8.
76. The research categories mentioned in the report optimistically included fisheries studies and research using submersibles and even research buoys. Burke, 11–12.
77. "Scientists Seek to Claim Cobb Seamount for U.S.," *Oceanology International* 3 (1968): 17.
78. Stewart Riley, "The Legal Implications of the Sea Use Program," *Marine Technology Society Journal*, February 1970, 31.
79. Frances and Walter Scott, *Exploring Ocean Frontiers* (New York: Parents' Magazine Press, 1970), 160.
80. At the 1958 Geneva Convention on the High Seas, the Soviets had pushed for peaceful development of the seafloor, and Butler interpreted this to indicate that the Soviets were primarily concerned with thwarting an American advantage. Ten years later, a Soviet memorandum proposed an international prohibition of fixed installations on the seabed and of "any other activities of a military character." William E. Butler, *The Soviet Union and the Law of the Sea* (Baltimore: Johns Hopkins University Press, 1971), 158.
81. A list of underwater habitats and their country of origin spanning 1962 to 1983 appears in James W. Miller and Ian G. Koblick, *Living and Working in the Sea* (New York: Van Nostrand Reinhold, 1984), 384–395.
82. *Civil Manned Undersea Activity: An Assessment* (Washington, DC: National Academy of Sciences and National Academy of Engineering, 1973), 18.
83. Scientists too were concerned that the Soviets had gained a research-platform advantage. When the Soviets fitted their submersible the *Severyanka* for fisheries research, *New Scientist* lamented that their submarine was "a matter for envy to oceanographers of other countries." "Submarine for Undersea Research," *New Scientist*, 5 February 1959, 272.
84. I refer here to Project Horizon, described in Frederick I. Ordway III, Mitchell R. Sharpe, and Ronald C. Wakeford, "Project Horizon: An Early Study of a Lunar Outpost," *Acta Astronautica* 17, no. 10 (1988): 1105–1121.
85. Austin, "Manned Undersea Structures," 35. *Popular Science* reported that Austin also forecast that undersea geothermal vents might eventually be tapped to "yield electricity, drinking water, and oxygen . . . for vast sea-floor colonies and secret submarine bases." William M. Holden, "Underground Steam," *Popular Science*, November 1968, 200.
86. Office of the Oceanographer of the Navy, *The Ocean Engineering Program of the U.S. Navy: Accomplishments and Prospects* (Alexandria, VA: Office of the Oceanographer of the Navy, 1967), 22. This document also includes artist impressions of several underwater habitat designs.
87. N. V. Breckner, "The Navy's Role in the Exploitation of the Ocean (Project Blue Water): Phase I," *Institute of Naval Studies*, 15 March 1968, 24.
88. Breckner, 25.

89. Harry Schwartz, "Science: To Conquer 'Inner Space,'" *New York Times*, 24 August 1969.
90. "Magnuson Is Named 'Conservationist of Year,'" *Statesville (NC) Record and Landmark*, 11 March 1976.
91. Advancement of Marine Sciences: Marine Sciences and Research Act of 1961, 87th Cong., 1st sess., Report No. 426 (Washington, DC: Government Publishing Office, 1961), 3–4. A comparison can be made to the anti-Soviet rationale used in establishing the Advisory Committee on Weather Control during the 1950s. See James Roger Fleming, *Fixing the Sky: The Checkered History of Weather Control* (New York: Columbia University Press, 2010), 176.
92. "Magnuson Sounds Stern Warning," *Port Angeles (WA) Evening News*, 31 July 1966.
93. Richard Fleming, then director of the University of Washington's Oceanography Department, noted in a 1963 letter to Rear Admiral E. C. Stephan, chairman of the navy's Deep Submergence Systems Review Group, that a project for the development of "an open-sea submerged buoy, with three-point moor and surface piercing mast, for use on the continental shelf," might interest the navy as a navigational aid, useful for underwater salvage operations. Richard Fleming Papers, folder 2, draft letter in response to a May 9 letter to Fleming from Stephan, University of Washington Library, Special Collections.
94. "Seamount Is Studied," *Daily Chronicle* (Centralia, WA), 29 October 1960, 7.
95. Hueter to Fleming, 11 March 1960, Richard Fleming Papers, University of Washington Library, Special Collections.
96. The story of the radio tower, the TOTEM buoy, is in Louise Burt and Miriam Ludwig, *Oceanography at Oregon State University: The First Two Decades, 1954–1975* (Corvallis: Oregon State University, 1998), 73. The voyage to the seamount aboard the *Brown Bear* also served as an experimental trial of new instrumentation. With the help of Boeing engineers, instruments were installed that automatically collected air and water temperatures, wind speed and direction. These were transmitted directly to a shore installation where they were stored in data processing equipment.
97. Hill Williams, "Oceanographic Institute Going Under. Nonprofit Group Loses Funding," *Seattle Times*, 20 September 1981, 47.
98. Gordon Sandison, "Sandison Sees Ocean Study Need," *Port Angeles (WA) Evening News*, 2 February 1967.
99. On 11 March 1960, T. F. Hueter, manager of Honeywell's Seattle-based development laboratory, wrote Richard Fleming, director of the University of Washington's Department of Oceanography, "I hope to be able to explore with you and Dr. Paquette further the ways and means in which Honeywell, through its Deep Sea Research Unit, can contribute toward a successful attack on the Sea Mount task by bringing to bear Honeywell's experience in acoustics, instrumentation and telemetering." Richard Fleming Papers, University of Washington Library, Special Collections. Prior to joining Honeywell, Hueter had worked at MIT on developing medical ultrasound techniques.

100. Oceanographic Institute of Washington, *Development of a Fixed Permanent Mooring System on the Pinnacle of Cobb Seamount* (Seattle: Oceanographic Institute of Washington, 25 April 1971), ii. https://apps.dtic.mil/dtic/tr/fulltext/u2/723242.pdf.
101. Cong. Rec. Senate, Vol. 114, Part 11, 14919, 24 May 1968. For more on Magnuson's lobbying for oceanographic work, see Jacob Hamblin, *Oceanographers and the Cold War: Disciples of Marine Science* (Seattle: University of Washington Press, 2005), 148.
102. "Demo Urges Exploration of Volcano," *Eugene (OR) Register Guard*, 26 May 1968.
103. John P. Craven, "Sea Power and the Sea Bed," *Proceedings Magazine* 92 (April 1966): 48–49.
104. See Christopher Drew, Sherry Sontag, and Annette Lawrence Drew, *Blind Man's Bluff: The Untold Story of American Submarine Espionage* (New York: Harper Collins, 2000).
105. Fifth meeting of the Oceanographic Commission of Washington, 24 April 1968, Agenda Item No. 5, Oceanographic Commission of Washington Records, University of Washington Library, Special Collections.
106. Tektites are glasslike mineral objects found on land and in the ocean, formed during meteorite impacts. The name was meant to underscore the connection of the mission to space exploration.
107. John Noble Wilford, "4 Scientists Will Live, Work for 60 Days on Floor of Ocean," *New York Times*, 2 May 1968.
108. The essential part played by such tinkerers has largely been overlooked by historians of postwar marine science; however, the critical importance of tinkering as a form of scientific work, particularly of science in the field, has been highlighted elsewhere. See, for example, Helen Rozwadowski, "'Simple Enough to Be Carried Out on Board': The Maritime Environment and Technological Innovation in the Nineteenth Century," in *Teknikens Landskap: En teknikhistorisk antologi tillägnad Svante Lindqvist*, ed. Ulf Larsson and Marika Hedin (Stockholm: Atlantis, 1998), 83–98; Simon Werrett, *Thrifty Science: Making the Most of Materials in the History of Experiment* (Chicago: University of Chicago Press, 2019); and Frank Nutch, "Gadgets, Gizmos, and Instruments: Science for the Tinkering," *Science, Technology, and Human Values* 21, no. 2 (Spring 1996): 214–228.
109. *Hearings before the Subcommittee on Oceanography of the Committee on Merchant Marine and Fisheries, House of Representatives*, Serial No. 91-5 (statements of Richard Waller, Conrad Mahnken, John VanDerwalker, Bureau of Commercial Fisheries; and Dr. Edward Clifton, Geological Survey, Department of the Interior) (Washington: U.S. Government Printing Office, 1969), 213–214.
110. Cheryl Wilson, "V.I. May Be Site of Another *Tektite* Program," *Virgin Islands Daily News*, 27 June 1969, 1, 8. Tektite II also took place in the Virgin Islands and ran for ten missions in 1970—including an all-female expedition team led by Sylvia Earle. It was then shipped to San Francisco, purchased by a nonprofit, and refurbished, but it was never again used as a habitat. PRINUL, the Puerto Rico International Undersea Laboratory, was later marketed as an extension of the Tektite program. See "*Tektite* Lives On," *Popular Science*, February 1972, 63.

111. W. E. Hoffmann and M. C. Hironaka, *Preliminary Site Survey for Ocean Construction Experiments*, Technical Report R-633 (Port Hueneme, CA: Naval Civil Engineering Laboratory, June 1969). The report notes that habitat programs, including Project Sea Use on Cobb Seamount, were "in conceptual stages."
112. The Makai Test Range of the Oceanic Foundation in Hawaii was a nonprofit organization operated as part of a complex of marine organizations that included the commercial Sea Life Park and the nonprofit Oceanis Institute, headed by Tap Pryor. An independently wealthy self-described marine biologist, Pryor served as a member of the Stratton commission, formed by Congress in 1966 and charged with making recommendations for the "full and wise use" of the marine environment. Pryor was appointed to the commission's panel on marine engineering and technology. David M. Karl, *UH and the Sea: The Emergence of Marine Expeditionary Research and Oceanography as a Field of Study at the University of Hawaii at Manoa* (Manoa: University of Hawaii, 2004), 5-2–5-4. See also, "Frontiersman of the Sea: Tap Pryor, Crusading Biologist," *Life Magazine*, 27 October 1967, 45–50.
113. Ralph Weiskittel, "Oceanography a Natural for Hawaii Development," *Cincinnati Enquirer*, 27 September 1969, 17. Seven different private companies backed the development at an initial cost of two million dollars. Wallace Mitchell, "$2 Million Undersea Test Range to Be Built off Makapuu Point," *Honolulu Advertiser*, 29 May 1966, A-13. It became clear that the greatest financial return rested on the promise of future research and development for the offshore oil industry. Purely academic projects would prove more difficult to fund.
114. OCW meeting minutes, 31 July 1968. Records of the Oceanographic Commission of Washington, University of Washington Library, Special Collections.
115. Battelle Memorial Institute, *Project Sea Use: Proposed Exploration of Cobb Seamount to the Oceanographic Commission of Washington* (Seattle: Battelle Memorial Institute, 24 April 1968), III-10.
116. Marti Mueller, "Oceanography: Who Will Control Cobb Seamount?," *Science* 161, no. 3838 (19 July 1968): 252–253.
117. Hill Williams, "On Pacific's Cobb Seamount," *Seattle Times* clipping, unknown date, courtesy of Spence Campbell. Rear Admiral Emory Day Stanley (1913–1999) graduated from the US Naval Academy in 1935. He served in the Pacific, Atlantic, and Mediterranean before retiring in 1965—at which point he moved to Seattle and became secretary of the Sea Use Council. In 1966 he founded Stanley Associates, a "government information technology firm." His 1935 Naval Academy yearbook jokingly describes him as "one of the most pro-Navy men since Phormio." Phormio was a celebrated Athenian naval commander who lived during the 5th century BC and fought in the Peloponnesian War.
118. "Minitat Possible in Puget Sound," *Daily Chronicle* (Seattle), 5 August 1970. Sharon Dodge was a Sea Use diver in the summer of 1974.
119. Mueller, "Oceanography," 252–253.
120. Stephan to Fleming, 9 May 1963, and Fleming's draft response. Richard Fleming Papers, University of Washington Library, Special Collections.

121. OCW meeting minutes, 31 July 1968. Records of the Oceanographic Commission of Washington, University of Washington Library, Special Collections.
122. Spence Campbell, *Lewis and Clark and Me* (Rochester, WA: Lewis and Clark Publishing, 2006).
123. These experiments are described in Spence Campbell, *After the Swim* (self-pub., CreateSpace, 2017).
124. Sheep were a good analogue human test organism because of the similarity of their vascular structure, and pigs were good analogue organisms for human bone vasculature—important for research on aseptic bone necrosis, a malady common in commercial divers.
125. Hill Williams, "Divers Spellbound by Seamount," *Seattle Times* clipping, unknown date, courtesy of Spence Campbell.
126. Author interview with Spence Campbell, 18 March 2016.
127. Campbell, interview.
128. Author interview with Roland White, 18 March 2016.
129. A complete roster of the Sea Use divers is in Bill Brubaker, *Seamount* (self-pub., CreateSpace, 2016), 145–146. Bill Brubaker is a retired Seattle-based journalist who served on the *Sea Use* Committee as the project historian. Divers of the first mission included Spence Campbell, Carl Eurick, Chuck Blackstock, Jim Gavin, Peter Taylor, and Dale Kister.
130. Spence Campbell scrapbook (2010), 15.
131. Biological work was conducted by Chuck Birkeland. Charles Birkeland, "Biological Observations on Cobb," *Northwest Science* 45, no. 3 (1971): 193–199.
132. Brubaker, *Seamount*, 91–100.
133. Robert Sheats was a World War II veteran. Captured by the Japanese in the Philippines, he survived the Bataan death march and several years in a POW camp. While a prisoner he was forced to work as a salvage diver. He later documented his experience in a memoir: Robert Sheats, *One Man's War: Diving as a Guest of the Emperor 1942* (Flagstaff, AZ: Best Publishing, 1998).
134. Spence Campbell scrapbook (2010), 121.
135. The most expensive piece of equipment was an atomic-powered naval transducer installed in the summer of 1973. This $10,000 instrument, powered by strontium-90 and designed to operate for up to ten years underwater, put out a low-frequency ultrasonic signal that could be picked up at a receiving station near San Diego. Atomic Energy Commission, "Draft Technical Progress Review," *Isotopes and Radiation Technology*, ed. Harold Atkinson and Jack Kahn 8, no. 1 (6 April 1970): 246. The exact purpose of this transducer, the details of which were not shared with the Sea Use aquanauts, likely related to the navy's submarine detection program. During the 1960s, Project Artemis, in the Atlantic, was an active transducer array system that also relied on infrastructure installed on the seabed. Argus Island was an oceanographic research platform built by the navy in 1961 off the coast of Bermuda. For another example of the importance of oceanography for submarine detection, see Simone Turchetti, "Sword, Shield and Buoys: A History of the NATO Sub-committee on Oceanographic Research, 1959–1973," *Centaurus* 54 (2012): 205–231.

136. Kennedy's inaugural address acknowledged multiple frontiers: "Together let us explore the stars, conquer the deserts, eradicate disease, tap the ocean depths." John F. Kennedy, "Inaugural Address," John F. Kennedy Presidential Library and Museum, http://www.jfklibrary.org/learn/about-jfk/historic-speeches/inaugural-address.
137. John F. Kennedy, "Address at Rice University on the nation's space effort," John F. Kennedy Presidential Library and Museum, https://www.jfklibrary.org/learn/about-jfk/historic-speeches/address-at-rice-university-on-the-nations-space-effort.
138. Andrew Scutro, "USN's NR-1 Wraps Up 40-year Career," *Defense News*, 8 December 2008.
139. Nicholas Flemming, "Man under the Sea," *New Scientist*, 3 December 1970, 366–368. Other articles in the same issue lamented the lack of progress in aquaculture and deep-sea mining.
140. "Tests Boost Deep Diving Hopes," *Palm Beach Post*, 1 October 1967. The patent for the ADS I, granted to Hugh D. Wilson, was filed in 1965. See "Device Involves Deep Sea Divers," *Nashua Telegraph*, 1 December 1965. The US Navy began using the ADS IV before developing its own, the Deep Dive System Mark I (DDS-Mk1), in 1968.
141. Eric Conway, "Bringing NASA Back to Earth: A Search for Relevance during the Cold War," in *Science and Technology during the Global Cold War*, ed. Naomi Oreskes and John Krige (Cambridge, MA: MIT Press, 2014), 256.
142. "US Plans Mobile Underwater Research Habitat," *New Scientist*, 3 November 1977, 273. NASA's Skylab was launched in 1973 and finally terminated in early July 1979.
143. "$21.5 Million Lab Is Being Designed for Oceanic Studies," *New York Times*, 24 July 1977. Ultimately, disagreement over proposed research applications, and the exorbitant expenses entailed, resulted in a redirecting of Oceanlab funds to existing research infrastructure. Hellwarth, *Sealab*, 255–256. Nevertheless, biological oceanographer G. Richard Harbison still hoped in 1983 that an Oceanlab-type "undersea research vessel" would soon be built. G. Richard Harbison, "The Structure of Planktonic Communities," in *Oceanography: The Present and Future*, ed. Peter G. Brewer (Woods Hole, MA: Woods Hole Oceanographic Institute, 1983), 31–33. See also Raymond Ramsay, "Oceanlab—Hindsight for Foresight," *Sea Technology* 35, no. 7 (1994): 85. Oceanlab Project case files, diagrams, and blueprints are in the records of the NOAA Undersea Research Program, Special Research Program Office, in the National Archives, Washington, DC.
144. Astronauts continue to train underwater, often living in Aquarius, the only remaining scientific underwater habitat in the United States.
145. Tom Parfitt, "Russia Plants Flag on North Pole Seabed," *Guardian* (UK), 2 August 2007.
146. William J. Broad, "China Explores a Frontier 2 Miles Deep," *New York Times*, 11 September 2010.
147. "China Is Planning a Massive Sea Lab 10,000 Feet Underwater," *Bloomberg News*, 7 June 2016, http://www.bloomberg.com/news/articles/2016-06-07/china-pushes-plan-for-oceanic-space-station-in-south-china-sea.

148. David Hambling, "Why Is Russia Sending Robotic Submarines to the Arctic?," *BBC Future*, 21 November 2017, http://www.bbc.com/future/story/20171121-why-russia-is-sending-robotic-submarines-to-the-arctic.
149. Douglas R. Burnett and Lionel Carter, "Deep Sea Observatories and International Law," in *Oceans 2009—Europe, Conference Proceedings*, Institute of Electrical and Electronics Engineers (2 October 2009), https://ieeexplore.ieee.org/document/5278102.

5. Ocean Science and Governance in the Anthropocene

Epigraph: Elisabeth Mann Borgese, *The Future of the Oceans: A Report to the Club of Rome* (Montreal: Harvest House, 1985), back cover.

1. Richard A. Frank, "The Law at Sea," *New York Times*, 18 May 1975.
2. Daniel C. Dunn, Caroline Jablonicky, Guillermo O. Crespo, Douglas J. McCauley, David A. Kroodsma, Kristina Boerder, Kristina M. Gjerde, and Patrick N. Halpin, "Empowering High Seas Governance with Satellite Vessel Tracking Data," *Fish and Fisheries*, 19 April 2018, 729–739.
3. Food and Agriculture Organization, *The State of the World Fisheries and Aquaculture: Contributing to Food Security and Nutrition for All* (Rome: United Nations, 2016).
4. See, for example, *Shifting Baselines: The Past and the Future of Ocean Fisheries*, ed. Jeremy B. C. Jackson, Karen E. Alexander, and Eric Sala (Washington: Island Press, 2011); W. Jeffrey Bolster, *The Mortal Sea: Fishing the Atlantic in the Age of Sail* (Cambridge, MA: Harvard University Press, 2012).
5. Furthermore, a 2008 study found that 35 percent of fish surveyed in the North Pacific had plastics in their stomachs. Christiana Boerger, Gwendolyn L. Lattin, Shelly L. Moore, Charles J. Moore, "Plastic Ingestion by Planktivorous Fishes in the North Pacific Central Gyre," *Marine Pollution Bulletin* 60, no. 12 (December 2010): 2275–2278. Microplastics are taken up by commercially harvested fish and shellfish and eventually consumed by humans.
6. Daniel Pauly, "Anecdoted and the Shifting Baseline Syndrome of Fisheries," *Trends in Ecology & Evolution* 10, no. 10 (October 1995): 430; Pauly, "Aquacalypse Now," *New Republic*, 27 September 2009.
7. Raymond Rogers, *The Oceans Are Emptying: Fish Wars and Sustainability* (New York: Black Rose Books, 1995), 84–85.
8. Bolster, *Mortal Sea*, 271–272.
9. David A. Kroodsma et al., "Tracking the Global Footprint of Fisheries," *Science* 359, no. 6378 (23 February 2018): 904–908.
10. Chiefly by Taiwan, South Korea, Spain, and China. David Tickler, Jessica J. Meeuwig, Maria-Lourdes Palomares, Daniel Pauly, and Dirk Zeller, "Far from Home: Distance Patterns of Global Fishing Fleets," *Science Advances* 4, no. 8 (1 August 2018): eaar3279.
11. Enric Sala, Juan Mayorga, Christopher Costello, David Kroodsma, Maria L. D. Palomares, Daniel Pauly, U. Rashid Sumaila, Dirk Zeller, "The Economics of Fishing the High Seas," *Science Advances* 4, no. 6 (6 June 2018): eaat2504.

12. "Group Wants to Really 'Sock It to' AEC on Safety Issue," *Daily Tribune* (Wisconsin Rapids, WS), 26 October 1974. Other commercial nuclear-powered vessels were also deemed failures, notably the American-built NS *Savannah* (launched 1959) and the German-built *Otto Hahn* (launched 1964).
13. Walter C. Patterson, *Nuclear Power* (Harmondsworth, UK: Penguin, 1976), 210.
14. Albert I, "Sur le Gulf-Stream. Recherches pour établir ses rapports avec la côte de France. Campagne de L'Hirondelle 1885" (Paris: Gauthier-Villars, Imprimeur-Libraire du Bureau des Longitudes, de l'École Polytechnique, 1886), 8.
15. Matthew Fontaine Maury, *The Physical Geography of the Sea* (New York: Harper & Brothers, 1858), 25.
16. Henry Stommel, *The Gulf Stream: A Physical and Dynamical Description* (Berkeley: University of California Press, 1958), ix. See also Jennifer Stone Gaines and Anne D. Halpin, "The Art, Music and Oceanography of Fritz Fuglister," Summer 2011, WHOI History Files, Woods Hole Oceanographic Institution Library and Archives.
17. Carroll Livingston Riker, *Power and Control of the Gulf Stream* (New York: Baker & Taylor, 1912), 9–10.
18. Robert G. Skerrett, "Warming the North Atlantic Coast," *Technical World Magazine*, March 1913, 62.
19. "To Move the Earth and Melt the Pole Dream of Engineer," *The Vancouver (BC) Sun*, 15 October 1912.
20. In the United States, Maury's *The Physical Geography of the Sea*, first published in 1855, long remained the most comprehensive compendium about ocean currents. Further detailed study of the Gulf Stream had been carried out in the 1890s by the American naval lieutenant John Elliot Pillsbury, commissioned to oversee Gulf Stream investigations of the US Coast and Geodetic Survey steamer *Blake*.
21. James Roger Fleming, *Fixing the Sky: The Checkered History of Weather Control* (New York: Columbia University Press, 2010), 201. The commission formation was part of a bill, H.R. 1779, 63rd Congress, first session, which failed because it lacked the support of Josephus Daniels, secretary of the navy, who deemed the project's ultimate goal unfeasible.
22. Carroll Livingston Riker, *International Police of the Seas* (New York: Baker & Taylor, 1915), 16–17.
23. "British Assured We Can't Swipe Gulf Stream," *Los Angeles Times*, 11 November 1949.
24. For instance, scientific consensus has not been reached to explain the separation of the Gulf Stream from the western boundary current just north of Cape Hatteras.
25. See Ronald Rainger, "'A Wonderful Oceanographic Tool': The Atomic Bomb, Radioactivity and the Development of American Oceanography," in *The Machine in Neptune's Garden: Historical Perspectives on Technology and the Marine Environment*, ed. Helen Rozwadowski and David K. van Keuren (Sagamore Beach, MA: Science History, 2004), 93–131.
26. See, for instance, the work of the NASA-funded Estimating the Circulation and Climate of the Ocean consortium.

27. The disappearance of sea ice can be equally disruptive. As sea ice recedes owing to a warming climate, denser deep water mixes higher in the water column, increasing salinity. Sigrid Lind, Randi B. Ingvaldsen, and Tore Furevik, "Arctic Warming Hotspot in the Northern Barents Sea Linked to Declining Sea-Ice Import," *Nature Climate Change* 8 (2018): 634–639.
28. M. A. Srokosz and H. L. Bryden, "Observing the Atlantic Meridional Overturning Circulation Yields a Decade of Inevitable Surprises," *Science* 348, no. 6241 (19 June 2015), http://science.sciencemag.org/content/348/6241/1255575.
29. Xianyao Chen and Ka-Kit Tung, "Global Surface Warming Enhanced by Weak Atlantic Overturning Circulation," *Nature* 559 (2018): 387–391. Additionally, we still have much to learn about how changes to the global climate in the past may have altered oceanic circulation patterns. See, for example, Jianghui Du, Brian A. Haley, Alan C. Mix, Maureen H. Walczak, and Summer K. Praetorius, "Flushing of the Deep Pacific Ocean and the Deglacial Rise of Atmospheric CO_2 Concentrations," *Nature Geoscience* 11 (2018): 749–755.
30. Elisabeth Mann Borgese, *The Oceanic Circle: Governing the Seas as a Global Resource* (New York: United Nations University Press, 1998).
31. Prue Taylor, "The Common Heritage of Mankind: Expanding the Oceanic Circle," *The Future of Ocean Governance and Capacity Development: Essays in Honor of Elisabeth Mann Borgese (1918–2002)*, ed. by International Ocean Institute—Canada (Leiden: Brill Nijhoff, 2018), 142.
32. Gail Jennes, "We Came from the Sea: Now, Says Thomas Mann's Daughter Elisabeth, We Must Return to It," *People* 14, no. 11 (15 September 1980).
33. Elisabeth Mann Borgese, *Mit den Meeren leben* (Cologne, Germany: Kiepenhauer & Witsch, 1999), 17–19. Translated in Agustín Blanco-Bazán, "Peace and the Law of the Sea," *Ocean Yearbook 18*, ed. by Aldo Chircop and Moiro McConnell (Chicago: University of Chicago Press, 2004): 88–89.
34. Committee to Frame a World Constitution, *The Preliminary Draft of a World Constitution* (Chicago: University of Chicago Press, 1948), 43.
35. Borgese, *Mit den Meeren leben*, 19.
36. Frank Hughes, "World State's Super-Secret Constitution! Plan Sponsored by Hutchins Bared," *Chicago Tribune*, 17 November 1947.
37. Ely Culbertson, "Book Reviews: The Preliminary Draft of a World Constitution," *Indiana Law Journal* 24 (1949): 482.
38. Quoted in Sunil M. Shastri, "Elisabeth Mann Borgese: A Life Dedicated to Pacem in Maribus," *Ocean Yearbook 18*, 78.
39. The founding date of the CSDI given by Borgese's biographer, Kerstin Holzer, is 1952; however, I could find no record of the center prior to 1959. Kerstin Holzer, *Elisabeth Mann Borgese: Ein Lebensportrait* (Berlin: Kindler, 2001), 173.
40. The term was borrowed from Pope John XXIII's 1963 encyclical, http://w2.vatican.va/content/john-xxiii/en/encyclicals/documents/hf_j-xxiii_enc_11041963_pacem.html.
41. W. Patrick McCray, *The Visioneers: How a Group of Elite Scientists Pursued Space Colonies, Nanotechnologies, and a Limitless Future* (Princeton, NJ: Princeton University Press, 2013), 5.

42. Graham M. Turner, "A Comparison of the Limits to Growth with 30 Years of Reality," *Global Environmental Change* 18, no. 3 (August 2008): 397–411.
43. Jacob Darwin Hamblin, *Poison in the Well: Radioactive Waste in the Oceans at the Dawn of the Nuclear Age* (New Brunswick, NJ: Rutgers University Press, 2008), 223.
44. Richard Petrow, *The Black Tide: In the Wake of Torrey Canyon* (London: Hodder & Stoughton, 1968); Robert Easton, *Black Tide: The Santa Barbara Oil Spill and Its Consequences* (New York: Delacorte, 1972).
45. More evidence for the continuing public fascination with marine mammal intelligence can be found in the 1973 film *Day of the Dolphin* and the much later 1986 film *Star Trek IV: The Voyage Home*.
46. For Cousteau as environmental advocate, see Gary Kroll, *America's Ocean Wilderness: A Cultural History of Twentieth-Century Exploration* (Lawrence: University of Kansas Press, 2008). For the history of Greenpeace, see Rex Weyley, *Greenpeace: How a Group of Ecologists, Journalists and Visionaries Changed the World* (Vancouver, BC: Raincoast Books, 2004).
47. Sunil M. Shastri, "Elisabeth Mann Borgese: A Life Dedicated to Pacem in Maribus," 79.
48. UNCLOS came into force in 1994. The United States, notably, as of 2019 has yet to ratify the agreement. The ultimate result was a more limited application of the legal concept of the common heritage of humankind than initially envisioned by Pardo, although the concept has since been applied in other international legal contexts such as human genomics and internet regulation. See Jean Buttigieg, "The Common Heritage of Mankind: From the Law of the Sea to the Human Genome and Cyberspace," *Symposia Melitensia: Adaptations* 8 (2012): 81–92.
49. Arvid Pardo, "First Statement to the First Committee of the General Assembly, November 1, 1967," *The Common Heritage: Selected Papers on Oceans and World Order 1967–1974*, International Ocean Institute Occasional Papers, no. 3 (Malta: Malta University Press, 1975), 2.
50. Holzer, *Elisabeth Mann Borgese*, 178.
51. Elisabeth Mann Borgese, "The Sea and the Dreams of Man," *Impact of Science on Society*, nos. 3/4 (1983): 480. She took up elements of this work again in Borgese, *The Oceanic Circle*, 56–58.
52. Borgese, "The Sea and the Dreams of Man," 488.
53. Borgese, 489.
54. Anne-Flore Laloë, "'Plenty of Weeds & Penguins': Charting Oceanic Knowledge," *Water Worlds: Human Geographies of the Ocean*, ed. Jon Anderson and Kimberley Peters (Surrey, UK: Ashgate, 2014), 40–42.
55. Allan C. Fisher Jr., "San Diego: California's Plymouth Rock," *National Geographic* 136 (July 1969): 145.
56. H. William Menard, *Marine Geology of the Pacific* (New York: McGraw-Hill, 1964), x.
57. Walter Munk, "Oceanography Before, and After, the Advent of Satellites," in *Satellites, Oceanography and Society*, ed. David Halpern (Amsterdam: Elsevier, 2000), 1–3.

58. Henry Stommel and E. D. Goldberg, "Oceanography: An International Laboratory," *Science* 165 (1969): 751.
59. Nelson G. Hogg and Rui Xin Huang, eds., *Collected Works of Henry Stommel* (Boston: American Meteorological Society, 1995), vol. 1: 221. Stommel wrote this in an unpublished editorial, "Statement about Priorities for Greenhouse Research," in 1990.
60. Robert Ballard, *The Eternal Darkness: A Personal History of Deep-Sea Exploration* (Princeton, NJ: Princeton University Press, 2000), 311.
61. The anthropologist Stefan Helmreich writes that "the sensory trajectory through which the deep sea has been scientifically apprehended has traveled from the tactile, to the auditory, to the visual." Stefan Helmreich, *Alien Ocean: Anthropological Voyages in Microbial Seas* (Berkeley: University of California Press, 2009), 35.
62. "Hacking the Ocean's Mysteries," *UW IT Newsletter*, August 2018, https://itconnect.uw.edu/hacking-the-oceans-mysteries/.
63. Experiments with adaptive robotics were carried out during the summer 2018 research cruise of the *Falkor* with funding from the Schmidt Ocean Institute based in Palo Alto, California.
64. Peter I. Macreadie et al., "Eyes in the Sea: Unlocking the Mysteries of the Ocean Using Industrial, Remotely Operated Vehicles," *Science of the Total Environment* 634 (September 2018): 1077–1091.
65. Adrienne Bernhard, "The Quest to Map the Mysteries of the Ocean Floor," *BBC Future*, 5 April 2018, http://www.bbc.com/future/story/20180404-the-quest-to-map-the-mysteries-of-the-ocean-floor.
66. Janet Davison, "Murky Waters," *CBC News*, 5 August 2018, https://newsinteractives.cbc.ca/longform/deep-sea-mining-environment.
67. I refer to the operations of the Canadian mining company Nautilus Minerals. See also John Hannigan, *The Geopolitics of Deep Oceans* (Malden, MA: Polity Press, 2016), 43–49.
68. Robert Blasiak, Jean-Baptiste Jouffray, Colette C. C. Wabnitz, Emma Sundström, and Henrik Österblom, "Corporate Control and Global Governance of Marine Genetic Resources," *Science Advances* 4, no. 6 (6 June 2018): eaar5237.
69. Andrew Merrie, "Commentary: Can Science Fiction Reimagine the Future of Global Development?," *Re.Think*, 1 June 2017, https://rethink.earth/can-science-fiction-reimagine-the-future-of-global-development/.
70. Emma Marris, "Bluetopia," *Nature* 550 (5 October 2017): 22–24. For a critique of the seasteading movement, see Philip E. Steinberg, Elizabeth Nyman, and Mauro J. Caraccioli, "Atlas Swam: Freedom, Capital, and Floating Sovereignties in the Seasteading Vision," *Antipode* 44, no. 4 (September 2012): 1532–1550.
71. David Titly, "The United States' Role in Maritime Transportation in the Arctic," *U.S. House of Representatives Committee Repository*, https://docs.house.gov/meetings/PW/PW07/20180607/108390/HHRG-115-PW07-Wstate-TitleyUSNRetD-20180607.pdf.
72. Peng Yining, "China Charting a New Course," *China Daily*, 20 April 2016.
73. Clive Schofield, "Dividing and Managing Increasingly International Waters: Delimiting the Bering Sea, Strait and Beyond," *Science, Technology, and New Chal-*

lenges to Ocean Law, ed. Harry N. Scheiber, James Kraska, and Moon-Sang Kwon (Leiden, Netherlands: Brill, 2015), 320.
74. Andrew E. Kramer, "The Nuclear Power Plant of the Future May Be Floating Near Russia," *New York Times*, 26 August 2018.
75. See Jeff Hecht, "Fiber-Optic Submarine Cables: Covering the Ocean Floor with Glass," in *Communications under the Seas: The Evolving Cable Network and Its Implications*, ed. Bernard Finn and Daqing Yang (Cambridge, MA: MIT Press, 2009), 45–58. Microsoft's Project Natick is one example of underwater data centers.
76. Mary Pols, "Bigelow Lab's 'Time Machine' Aims to Find Out How Shellfish Will Fare in the Future," *Portland (ME) Press Herald*, 29 July 2018.
77. Scientists working in certain branches of oceanographic work might debate this claim; exploration remains an important component of deep-sea research, and the mythos of an endless frontier has long justified marine research. See Helen Rozwadowski, "Arthur C. Clarke and the Limitations of the Ocean as a Frontier," *Environmental History* 17, no. 3 (2012): 578–602. See also Erik Dücker's analysis of the accelerated pace of habitat discovery. Erik Dücker, "News from an Inaccessible World: The History and Present Challenges of Deep-Sea Biology" (PhD diss., Radboud University, Netherlands, 2014), 146–147.
78. R. P. Dziak, J. H. Haxel, J.-K Lau, Sara Heimlich, Jacqueline Caplan-Auerbach, David Mellinger, Haru Matsumoto, and B. Mate, "A Pulsed-Air Model of Blue Whale B Call Vocalizations," *Scientific Reports* 7, no. 1 (22 August 2017), https://www.nature.com/articles/s41598-017-09423-7.
79. Human-caused sea-surface temperature increases can occur gradually or suddenly. One study found that 87 percent of marine heat waves "are attributable to human-induced warming, with this ratio increasing to nearly 100 per cent under any global warming scenario exceeding 2 degrees Celsius." Projection models suggest that future marine heat waves will push "marine organisms and ecosystems to the limits of their resilience and even beyond." Thomas L. Frölicher, Erich M. Fischer, and Nicolas Gruber, "Marine Heatwaves under Global Warming," *Nature* 560 (2018): 360.
80. Livia Albeck-Ripka, "Climate Change Brought a Lobster Boom. Now It Could Cause a Bust," *New York Times*, 21 June 2018.
81. Remarkably, scientists have discovered that human activities at sea can also affect marine climate on a very localized scale. The soot particulates released by container ships in densely concentrated shipping areas increase the quantity of lightning strikes (an indicator of storm intensity) by a factor of two. See Joel A. Thornton, Katrina S. Virts, Robert H. Holzworth, and Todd P. Mitchell, "Lightning Enhancements over Major Oceanic Shipping Lanes," *Geophysical Research Letters* 44, no. 17 (September 2017): 9102–9111.
82. Kevin Scott, Petra Harsanyi, Alastair R. Lyndon, "Understanding the Effects of Electromagnetic Field Emissions from Marine Renewable Energy Devices (MREDs) on the Commercially Important Edible Crab, *Cancer pagurus* (L.)," *Marine Pollution Bulletin* 131, Part A (June 2018), 580–588.
83. Kaela Slavik, Carsten Lemmen, Wenyan Zhang, Onur Kerimoglu, Knut Klingbeil, and Kai W. Wirtz, "The Large-Scale Impact of Offshore Wind Farm Structures

on Pelagic Primary Productivity in the Southern North Sea," *Hydrobiologia* (2018), https://doi.org/10.1007/s10750-018-3653-5.

84. Malin L. Pinsky, Gabriel Reygondeau, Richard Caddell, Juliano Palacios-Abrantes, Jessica Spijkers, William W. L. Cheung, "Preparing Ocean Governance for Species on the Move," *Science* 360, no. 6394 (15 June 2018): 1189–1191.

85. Roger Revelle and Hans Seuss, "Carbon Dioxide Exchange between the Atmosphere and Ocean and the Question of an Increase in Atmospheric CO_2 during the Past Decades," *Tellus* 9 (1957): 19–20. Revelle, who "liked great geophysical experiments," was less worried about the implications of global climate change and more concerned with establishing a program for systematic monitoring. See Joshua Howe, *Behind the Curve: Science and the Politics of Global Warming* (Seattle: University of Washington Press, 2014), 19.

86. Roger Harrabin, "'Coral Lab' Offers Acidity Insight," *BBC News*, 11 March 2009.

87. Elizabeth Kolbert, *The Sixth Extinction: An Unnatural History* (New York: Henry Holt, 2014), 111–124.

88. SCOR Working Group, "Geotraces: An International Study of the Marine Biogeochemical Cycles of Trace Elements and Their Isotopes," *Chemie der Erde* 67 (2007), 1.

89. A complete list of past and present cruises is provided on the Geotraces website. https://www.bodc.ac.uk/geotraces/cruises/programme/.

90. The meaning of "climate sink" is literal in a way, because plankton takes in carbon dioxide, then sinks as marine snow, trapping the carbon dioxide in deep-sea sediments.

91. For one example, see Philip W. Boyd et al., "A Mesoscale Phytoplankton Bloom in the Polar Southern Ocean Stimulated by Iron Fertilization," *Nature* 407 (12 October 2000): 695–702.

92. Hugh Powell, "Fertilizing the Ocean with Iron: Should We Add Iron to the Sea to Help Reduce Greenhouse Gases in the Air?," *Oceanus Magazine*, 13 November 2007, https://www.whoi.edu/oceanus/feature/fertilizing-the-ocean-with-iron/.

93. Powell.

94. Blooms may deplete oxygen at depth as bacteria decompose the plankton, consuming oxygen and releasing carbon dioxide back into the ocean. The addition of dissolved carbon dioxide into the water column contributes to ocean acidification because carbon dioxide bonds with water molecules to form carbonic acid.

95. Roger S. Barga and John Delaney, "Observing the Ocean—A 2020 Vision for Ocean Science," *The Fourth Paradigm: Data Intensive Scientific Discovery* (Seattle: Microsoft Research, 2009), 32.

96. See Eric Conway, "Drowning in Data: Satellite Oceanography and Information Overload in the Earth Sciences," *Historical Studies in the Physical and Biological Sciences* 37, no. 1 (September 2006): 127–151.

97. Mark Schrope, "Pirates Attack US Research Ship off Somalia," *Nature* 413, no. 97 (13 September 2001), 97.

98. Antony Adler, "The Ship as Laboratory: Making Space for Field Science at Sea," *Journal of the History of Biology* 47, no. 3 (2014): 333–362.
99. Information about the origins of the Argo program can be found online, http://www.argo.ucsd.edu/. See also Robert J. Fricke, "Down to the Sea in Robots," *Technology Review* 97, no. 7 (1994): 46–55.
100. John Delaney, "Wiring an Interactive Ocean," filmed April 2010 during Mission Blue Voyage, TED video, 20:43, https://www.ted.com/talks/john_delaney_wiring_an_interactive_ocean?utm_campaign=tedspread&utm_medium=referral&utm_source=tedcomshare.
101. William Yardley, "'Bringing the Ocean to the World,' in High-Def," *New York Times*, 4 September 2007.
102. Names of modern oceanographic projects frequently include "experiment." A few examples are the Mid-Oceans Dynamics Experiment, the Tasmania Internal Tide Experiment, and the North-East Pacific Time-Series Underwater Networked Experiments.
103. Robert F. Service, "Oceanography's Third Wave," *Science* 318, no. 5853 (16 November 2007): 1056–1058.
104. J. Emmett Duffy, "Ocean 2.0," in *Living in the Anthropocene: Earth in the Age of Humans*, ed. W. John Kress and Jeffery K. Stine (Washington: Smithsonian Books, 2017), 92.
105. Kimbra Cutlip, "Real Time Evidence Leads Government of Belize to Reverse Decision," *SkyTruth*, 26 October 2016, https://www.skytruth.org/2016/10/real-time-evidence-leads-government-of-belize-to-reverse-decision/.
106. Tryggvi Adalbjornsson, "A Victory for Coral: Unesco Removes Belize Reef from Its Endangered List," *New York Times*, 27 June 2018.
107. See Paul Dudley White, Samuel W. Matthews, and J. Baylor Roberts, "Hunting the Heartbeat of a Whale: A Scientific Expedition Sails a Lonely Mexican Lagoon in Search of Clues to the Mysteries of the Human Heart," *National Geographic* 110, no. 1 (July 1956): 49.
108. James Cheshire and Oliver Uberti, *Where the Animals Go: Tracking Wildlife with Technology in 50 Maps and Graphics* (New York: W. W. Norton, 2017), 85.
109. L. Parrott et al., "A Decision Support System to Assist the Sustainable Management of Navigation Activities in the St. Lawrence River Estuary, Canada," *Environmental Modelling and Software* 26, no. 12 (December 2011): 1403–1418.
110. "Nature is Speaking," Conservation International. 5 October 2014, https://www.conservation.org/nature-is-speaking/Pages/default.aspx.
111. William C. Hushing, "The Impact of High Performance Science and Technology on Manpower Requirements at the Undersea Interface," in *Critical Interfaces for Engineers and Scientists: 4 Appraisals* (New York: Engineers Joint Council, 1967), 2.
112. Rowan Jacobsen, "Obituary: Great Barrier Reef (25 Million BC–...)," *Outside Magazine*, 11 October 2016.
113. Chris D'Angelo, "Great Barrier Reef Obituary Goes Viral, to the Horror of Scientists," *Huffington Post*, 14 October 2016, https://www.huffingtonpost.com/entry/scientists-take-on-great-barrier-reef-obituary_us_57fff8f1e4b0162c043b068f.

114. Steve Katona, "Good News: The Ocean Isn't Dying," *Ocean Health Index*, 21 April 2015, http://www.oceanhealthindex.org/news/Ocean_Isn't_Dying.
115. Sheila Jasanoff has argued, "Scientific facts bearing on the global environment never take root in a neutral interpretive field; they are dropped into contexts that have already been conditioned to produce distinctive cultural responses to scientific claims." Sheila Jasanoff, "A New Climate for Society," *Theory, Culture and Society* 27 (2010): 240.
116. The Royal Canadian Mounted Police arrested a man who accused Cullen of acting on behalf of the nuclear power industry. Mark Hume, "Charges Laid against B.C. Man Who Called for Death of Fukushima Researcher," *Globe and Mail*, 6 November 2015.
117. Mark Hume, "Canadian Researcher Targeted by Hate Campaign Over Fukushima Findings," *Globe and Mail*, 1 November 2015.
118. Henry Fountain, "A Rogue Climate Experiment Outrages Scientists," *New York Times*, 18 October 2012.
119. See Kim Martini, "The Ocean Cleanup, Part 2: Technical Review of the Feasibility Study," *Deep Sea News*, 14 July 2014, http://www.deepseanews.com/2014/07/the-ocean-cleanup-part-2-technical-review-of-the-feasibility-study/.
120. For a discussion of the past and future impact of sea-level rise on human civilization, see Brian Fagan, *The Attacking Ocean: The Past, Present, and Future of Rising Sea Levels* (New York: Bloomsbury, 2013).
121. For the iconography, and limitations, of polar bears as symbols for effective climate change communication, see Dorothea Born, "Bearing Witness? Polar Bears as Icons for Climate Change Communication in *National Geographic*," *Environmental Communication*, February 2018, https://doi.org/10.1080/17524032.2018.1435557; Anna Westerstahl Stenport and Richard S. Vachula, "Polar Bears and Ice: Cultural Connotations of Arctic Environments That Contradict the Science of Climate Change," *Media, Culture and Society* 39, no. 2 (2017): 282–295.
122. For instance, a former head of the Environmental Protection Agency, Scott Pruitt, declared that it was "arrogant" to set target global temperature ranges for the future.
123. Nancy Knowlton, "Why Do We Have Trouble Talking About Success in Ocean Conservation?," *Smithsonian Magazine*, 12 June 2014, https://www.smithsonianmag.com/science-nature/why-do-we-have-trouble-talking-about-success-ocean-conservation-180951727/.
124. Elizabeth Kolbert, "Unnatural Selection: What Will It Take to Save the World's Reefs and Forests?," *New Yorker*, 18 April 2016, https://www.newyorker.com/magazine/2016/04/18/a-radical-attempt-to-save-the-reefs-and-forests.
125. Bruno Latour, *Politics of Nature: How to Bring the Sciences into Democracy* (Cambridge, MA: Harvard University Press, 2004), 61.
126. Erin L. O'Donnell and Julia Talbot-Jones, "Creating Legal Rights for Rivers: Lessons from Australia, New Zealand, and India," *Ecology and Society* 23, no. 1 (2018), https://doi.org/10.5751/ES-09854-230107.
127. Jean-Michel Cousteau, "An Exciting and Crucial Moment in History: Paris Climate Conference 2015," *Ocean Futures Society*, 3 December 2015, http://www.oceanfutures.org/news/blog/paris-climate-conference-2015.

128. Marine scientists have adopted the term "plastisphere" to refer to the ecosystem of marine organisms living on plastic debris. Kim De Wolff argues that such designations alter our interpretation of the boundary between the living and nonliving world. Kim De Wolff, "Plastic Naturecultures: Multispecies Ethnography and the Dangers of Separating Living from Nonliving Bodies," *Body and Society* 23, no. 3 (2017): 23–47.
129. An advocate of judicious harvesting is Ray Hilborn. See Ray Hilborn and Ulrike Hilborn, *Overfishing: What Everyone Needs to Know* (Oxford: Oxford University Press, 2012).
130. For examples of some of the possible negative outcomes resulting from marine climate engineering, see Alexander Proelss, "International Legal Challenges concerning Marine Scientific Research in the Era of Climate Change," in *Science, Technology, and New Challenges to Ocean Law*, ed. Harry N. Scheiber, James Kraska, and Moon-Sang Kwon (Leiden, Netherlands: Brill Nijhoff, 2015), 282.
131. Elspeth Probyn, *Eating the Ocean* (Durham, NC: Duke University Press, 2016), 5.
132. Sylvia A. Earle, *The World is Blue: How Our Fate and the Ocean's Are One* (Washington: National Geographic Society, 2009), 262.
133. Helen Rozwadowski writes, "Human motives ... matter as much as biological interactions or chemical reactions." Helen Rozwadowski, *Vast Expanses: A History of the Ocean* (London: Reaktion Books, 2018), 12.

Conclusion

Epigraph: Julien Moreton, *Life and Work in Newfoundland: Reminiscences of Thirteen Years Spent There* (London: Rivingtons, Waterloo Place, 1863), 40.

1. "Follow the actors" is the oft-repeated dictum of actor-network theory. Bruno Latour, *Reassembling the Social: An Introduction to Actor-Network-Theory* (Oxford: Oxford University Press, 2005).
2. For an example of this approach, see Samantha Muka, "Portrait of an Outsider: Class, Gender, and the Scientific Career of Isa M. Mellen," *Journal of the History of Biology* 47 (2014): 29–61. I address the work of marine station assistants in more detail in Antony Adler, "The Hybrid Shore: The Marine Station Movement and Scientific Uses of the Littoral, 1843–1910," in *Soundings and Crossings: Doing Science at Sea, 1800–1970*, ed. Katherine Anderson and Helen Rozwadowski (Sagamore Beach, MA: Science History, 2016), 147–178.
3. Steven Lee Myers, "With Ships and Missiles, China Is Ready to Challenge U.S. Navy in Pacific," *New York Times*, 29 August 2018.
4. Keyport, Washington, is the home of the navy's first drone squadron at the Naval Undersea Warfare Center.
5. Anne Gearan, "Clinton Announces New Ocean Panel," *ABC News* (no date given), https://abcnews.go.com/Technology/story?id=120046&page=1.
6. Madeleine Carlisle, "Trump's Offshore-Drilling Plan Is Roiling Coastal Elections," *Atlantic*, 5 August 2018, https://www.theatlantic.com/politics/archive/2018/08/trumps-offshore-drilling-plan-is-roiling-coastal-elections/566726/.

7. Bethan C. O'Leary, Marit Winther-Janson, John M. Bainbridge, Jemma Aitken, Julie P. Hawkins, and Callum M. Roberts, "Effective Coverage Targets for Ocean Protection," *Conservation Letters* 9, no. 6 (November–December 2016): 398–404.
8. More is still known about the seafloor and upper layers of the oceans than the midwater zone, two hundred to one thousand meters below the surface. Advances in remote sensing and robotics will make this region more accessible for study.
9. UNESCO, Media Services, "Intergovernmental Oceanographic Commission," http://www.unesco.org/new/en/media-services/single-view/news/towards_the_ocean_we_need_for_the_future_we_want_i/.
10. Sylvia Earle introduced this concept in a 2009 TED talk. "My Wish: Protect Our Oceans," Filmed February 2009 in Long beach, CA, TED video, 18:05, https://www.ted.com/talks/sylvia_earle_s_ted_prize_wish_to_protect_our_oceans.
11. Helena Pozniak, "The Blue Planet Effect: Why Marine Biology Courses Are Booming," *Guardian* (UK), 12 January 2018.
12. Erin Biba, "Awesome Jobs: Meet Julie Huber, Deep Sea Microbiologist," *Tested*, 12 December 2017, https://www.tested.com/science/804290-awesome-jobs-meet-julie-huber-deep-sea-microbiologist/. There are many examples of citizen marine-science projects. See, for example, Michael Bear, "Perspectives in Marine Citizen Science," *Journal of Microbiology and Biological Education* 17, no. 1 (March 2016): 56–59.
13. J. R. McNeill and Peter Engelke, *The Great Acceleration: An Environmental History of the Anthropocene since 1945* (Cambridge, MA: Harvard University Press, 2014), 209–210.
14. William Cronon, "The Uses of Environmental History," *Environmental History Review* 17, no. 3 (Autumn 1993): 5.
15. Nancy Langston, "Editor's Note," *Environmental History* 18, no. 1 (2013), 1.

Acknowledgments

I wish to give special thanks to Bruce Hevly, Simon Werrett, Ray Jonas, and Adam Warren in the Department of History at the University of Washington. I have learned so much from them over the years. In the UW School of Oceanography, John Baross, Jody Deming, Paul Johnson, Karl Banse, and Russ McDuff patiently entertained my questions and suggested important topics to pursue. The conversations I had with them greatly inform this project.

Eric Mills shared his teaching materials with me when I wrote to him in 2011. He was the first to encourage me to pursue research in the history of oceanography and has continued to provide guidance and sage advice. In taking on this field of study I have benefited from being part of a wonderful and supportive community of scholars working on the history of marine science. I am grateful to Keith Benson, Jacob Hamblin, Katharine Anderson, Jennifer Hubbard, Christine Keiner, Samantha Muka, Penelope Hardy, Kelly Bushnell, Erik Dücker, and Alistair Sponsel. In particular, I am indebted to Helen Rozwadowski for her generosity over the years. Her scholarship is a source of great inspiration.

Many people aided me with the archival work for this project. I wish to give special thanks to Jacqueline Carpine-Lancre, archivist at the royal palace of Monaco. Carpine-Lancre shared with me her extensive research notes and offered indispensable aid navigating the archives. Much of my writing on Prince Albert, and the history of oceanography in France, originated from my conversations with her. In Roscoff, France, André Toulmand and his wife, Claude, graciously hosted me during my time working in the marine station library. At the Scripps Institution of Oceanography Archives I was assisted by Heather Smedberg, and in the Special Collections of the Imperial College of London by Bryony Hooper. Chuck Blackstock and Spence Campbell were generous with their time and provided invaluable assistance when I contacted them to learn more about the history of Project Sea Use.

My research abroad was made possible through a fellowship from the American Geophysical Union, and a grant provided by the National Science Foundation. In 2015–2016 I was a research fellow in the Program on Science, Technology, and Society at Harvard University. The readings we shared and the conversations I had with scholars in that group were very helpful. I thank Sheila Jasanoff for inviting me to join that community. In Massachusetts, I was also supported through the Karush Library Fellowship at the Marine Biological Laboratories library in Woods Hole. At Carleton College I have found a welcoming and supportive community of colleagues.

Chapter 2 draws from work previously published in two chapters in two different edited volumes: "The Hybrid Shore: The Marine Station Movement and Scientific Uses of the Littoral, 1843–1910," in *Soundings and Crossings: Doing Science at Sea, 1800–1970*, ed. Katharine Anderson and Helen M. Rozwadowski (Sagamore Beach, MA: Science History, 2016), and "Legitimizing Marine Field Science: Albert I of Monaco and the Institutionalization of Oceanography," in *Understanding Field Science Institutions*, ed. Helena Ekerholm, Karl Grandin, Christer Nordlund, and Patience A. Schell (Sagamore Beach, MA: Science History Publications, 2017). I thank the editors of those works for their advice and guidance as I first began writing about these topics.

At Harvard University Press I want to thank my editor, Jeff Dean, for taking an interest in my work, for offering constructive critique on drafts, and for his patience along the way. I am also indebted to Emeralde Jensen-Roberts and Stephanie Vyce for their invaluable assistance. Sherry Gerstein saw this work through the final stages of production, and I am also extremely grateful to Mary Ann Short for her meticulous copyediting of my manuscript.

Finally, I owe my love of history to my mother, Judith Adler. As a scholar, I have benefited from our conversations and her insights. My godfather, Volker Meja, has been a source of support throughout my life. I also would not be completing this book were it not for the steadfast encouragement of my wife, Rika Anderson. I am so grateful for all that we share.

Index

Agassiz, Alexander, 33, 35, 75, 185n101, 192n43, 196n87, 200n3, 201n13
Aimé, Georges, 181n44
Akademik Lomonosov (floating nuclear reactor), 155
Albatross (USS), 5, 201n13
Albert I of Monaco, 9, 46–47, 60–73, 139, 166; critique of national expeditionary model, 64; death and legacy, 72–73; diplomacy, 66; education, and interest in paleontology, 61–62; and global international science, 70–73; and Museum of Natural History in Paris, 195n83; popularization of science, 65, 68; reaction to World War I, 71. *See also* Institut Océanographique; Musée Océanographique
Allen, Paul, 163
Andouin, Jean-Victor, 189n10
Animal tracking, 160–161. *See also* Whales
Anthropocene: oceans, 150–156, 161–165; science and governance in, 135–165
Aquarium (plural: aquaria), 26–30, 38, 43, 50, 55, 56; Boston Aquarial Gardens, 27–28; home and public aquaria, 26–30; Regent's Park Fish House, 27; as scientific instrument, 29, 191n22
Aquatic ape hypothesis, 105–106, 149
Arcachon exposition, 36–37
Arctic Waters Pollution Prevention Act, 114
Artificial intelligence. *See* Computer modeling
Artificial islands, 112–113, 134, 168
Atomic Energy Commission, 111, 138, 211n36
Atoms for Peace program, 138
Attenborough, David, 170
Austin, Carl F., 111, 118, 212

Baird, Spencer F., 40. *See also* US Fish Commission
Ballard, Robert, 132, 152
Banks, Joseph, 18
Barnum, P. T., 27–29
Bathythermograph, 94, 103, 208n5. *See also* Spilhaus, Athelstan
Battelle Memorial Institute, 121, 125, 127
Belize Barrier Reef, 160
Belknap, George, 34

Berlin Institute of Oceanography, 76
Big science, 11, 80; oceanography as, 45, 184n93
Biodiversity, threats to, 165, 168, 169
Blackstock, Chuck, 129, 232
Blake (USS), 5, 63, 221n20
Bolster, W. Jeffrey, 6, 42
Bonaparte, Roland, 70
Borgese, Elisabeth Mann, 135, 139, 144–149, 167; Chicago Committee, 145; "humanized ocean," 149; "Mother of the Oceans," 144; oceans as a political laboratory, 148; Pacem in Maribus, 147–148
Borgese, Giuseppe E., 145
Boyle, Robert, 14–17
Bryan, William Alanson, 85, 203n40
Buchanan, John Young, 64–65, 68, 196n95, 197n98, 198n113
Buckland, Frank, 41, 187n136
Bureau of Naval Weapons, 96
Bush, George W., 169
Bush, Vannevar, 93, 106
Butler, Henry D., 28

Calder, William Musgrave, 142
Caldwell Jr., Turner F., 125
Campbell, Spence, 127–131, 232. *See also* Virginia Mason Medical Center
Canada, 42, 82, 104, 144, 136–137, 141, 176n14, 202n35. *See also* Newfoundland
Carnegie Institution, 106
Carpenter, William B., 31–32
Carpine-Lancre, Jacqueline, 175n5, 198n108
Carson, Rachel, 146
Caullery, Maurice, 57
Cavarrubias, Miguel, 89
Center for the Study of Democratic Institutions (CSDI), 146
Challenger (H.M.S.), 5, 30–35, 41, 47, 176n10, 183n88, 184n93; collections, 33–34, 184n96; reports of the expedition, 35

Chalon, Jean, 51, 70
Chicago Committee to Frame a World Constitution, 145
Clarke, Arthur C., 135
Climate change, 144, 149–151, 154, 156–158, 164–165, 222n27, 225n81, 226n85; iron seeding, 158, 162–163; warming seas, 155–156
Clinton, Bill, 168
Club of Rome, 146–147
Coastal School of Deep Sea Diving, 127
Coast and Geodetic Survey, 24, 32, 47, 182n61, 221n20
Cobb Seamount, 117, 119–131; discovery of, 120; legal concerns regarding, 117, 126; occupation of, 126
Cod fishery collapse, 40, 137
Cold war, 11; competition over marine territory, 117–118; diplomacy and marine science, 103; and international cooperation in oceanography, 98
Commoner, Barry, 146
Computer modeling, 144, 146, 161
Conrad, Joseph, 25
Continental shelf laboratories, 111
Cook, James, 17–18, 180n19, 203n43
Coral, 16, 50, 83, 134, 147, 153, 160, 162, 164, 190n16
Coste, Victor, 49, 190n13
Coup, William Cameron, 28
Cousteau, Jacques, 49, 97, 104–110, 147, 210n24, 210n26, 210n28, 210n32, 211n34; *Homo aquaticus*, 106–107; Conshelf Habitat program, 107, 109, 110, 211n33; Mysterious Island platform, 109, 211n42
Cousteau, Jean-Michel, 164
Craven, John, 122
Crayford, John E., 104
Crinoid studies, 21. *See also* Sars, Michael
Cronon, William, 171
Crowdsourcing ocean data, 160
Cullen, Jay, 162, 228n116

Cutler, Leland, 91
Cuvier, Georges, 49, 189n10

Dall, William Healey, 34
Daly, Reginald Aldworth, 86
Dan, Katsuma, 92, 205n62
Darwin, Charles, 18, 21–22, 34, 55–56, 190n17
Deacon, Margaret, 33, 35, 176, 181n39
Deepwater Horizon, 169
Delage, Yves, 68, 70
Delaney, John, 150–160
Delcassé, Théophile, 58
Deloughrey, Elizabeth. *See* Oceanic turn
Department of Defense, 114, 213n73
Dodge, Sharon, 126, 217n118
Dohrn, Anton, 46–47, 50, 54–57
Dredge, 21–23, 29, 44, 55, 181n41, 184n91; dredging committee, 22, 191nn19–20; dredging papers, 22
Dreyfus, Alfred, 61
Drygalski, Erich Dagobert von, 68
Duffy, Emmett J., 160
Duncan, Peter Martin, 33–34
Dying seas narrative, 161–165

Earle, Sylvia, 165, 170, 216n110
Earth and Space Sciences Agency, 120
Edwards, Alphonse Milne, 31, 62, 189n10, 196n87
Ehrlich, Paul, 146
Eisenhower, Dwight, 138
Engelke, Peter, 170
Environmental collapse, 136, 165
Estai (fishing trawler). *See* Turbot war
Exclusive Economic Zones (EEZs), 112, 136, 156

Fabre-Domergue, Paul, 68
Fairtry (fishing trawler), 137
Falco, Robert, 107, 211n33
Falejczyk, Frank J., 108. *See also* Liquid-breathing experiments

Falla, R. A., 91
Federal Science Exhibit, 94–95, 97; science demonstrators, 96
Fields, David, 155
Findlay, John, 93
Fishing: aquaculture, 36–37, 39, 44; cod fishery, 39–40, 42, 137; commercial dredge, 184n91; factory trawling, 39–40, 71, 137, 186n128, 187n131, 187n133; global fish stocks, 136, 165; growth of industrialized fishing, 36–44; illegal fishing, 160; lobster fishery, 155; North Sea fishery, 57–58; royal fisheries commissions, 40; shifting baselines, 136
Fleming, James, 142
Fleming, Richard, 120, 126–127, 215n93, 215n99
FLIP platform, 109
Forbes, Edward, 22–23, 31, 166; azoic theory, 23, 31, 181n50
Ford, Harrison, 161
Ford Foundation, 146
Forster, Johann Reinhold, 17–18, 179n18, 180n22, 203n43
France: Brittany region, 40, 49, 51–52; Franco-Prussian War, 37, 44, 52; French and Prussian scientific rivalry, 9, 44, 52–53, 57
Frank, Richard A., 136
Franklin, Benjamin, 17, 139
Fraser, Charles McLean, 84–85
Frédéricq, Léon, 53
Fukushima, 162
Fulton, Charles, 113
Future, modern concepts of the, 2, 7–8

Gagnan, Émile, 104–105, 209n15
Gandhi, Mohandas, 144
Gates, Ruth D., 164–165
Gene patenting, 153
General Bathymetric Chart of the World (GEBCO), 70–72, 198n118
General Electric (company), 102, 108, 111

General Motors' Futurama, 110
Geotraces, 157
Germany, 33, 57–58, 64, 66, 70–71, 77, 142, 145, 153, 176n14, 198n115
Ghaleb, Osman, 57
Giard, Alfred, 54, 57
Goethe, Johann Wolfgang von, 56
Goldberg, E. D., 151
Golden age of global oceanography, 99
Golden Gate International Exposition, 10, 77, 79, 86–87, 91; Pacific House, 87–90
Gosse, Philip Henry, 29–30, 183n80
Grand Capri Republic, 113
Great Barrier Reef, 147, 162
Great International Fisheries Exposition, 37–38, 41–43, 186n121
Great ship narrative, 4
Greenpeace, 147
Gregory, Herbert, 78–79, 91
Groeben, Christiane, 3–4
Grotius, Hugo, 112
Gulf Stream, 17, 24, 63, 115, 139–144, 221n20, 221n24; and weather control, 141–142; drift experiments, 63; weaponizing of, 143. *See also* Riker, Carroll Livingston

Haeckel, Ernst, 33, 55
Halley, Edmund Halley, 15, 17
Hamblin, Jacob Darwin, 98, 177, 178
Hardy, Sir Alistair. *See* Aquatic ape hypothesis
Haverstock, Elizabeth, 77
Herdman, William, 47, 76, 181
Higgins, Henry H., 33
Hirondelle, 63, 196n86
Hirondelle II, 197n96, 200n133
Honeywell (company), 102, 120–121, 127, 129, 215n99
Hooke, Robert, 15
Huber, Julie, 170
Hushing, William C., 161
Hutchins, Robert M., 145–146

Huxley, Thomas Henry, 18–19, 34, 40–44, 180n24
Hydrographic Office, 25
Hydrography, 15, 19, 24–25, 81

Imagination: and the marine environment, 6, 13–14, 26, 29, 32, 104; and future predictions, 12, 102, 133, 153, 164; and the history of marine science, 166, 171–172
Institut für Meereskunde, 198n115
Institut Océanographique, 9, 68–69, 72, 196n87, 196n95, 197n103, 198n115
International Council for the Exploration of the Sea (ICES), 10, 44, 58, 80, 84, 115; French absence from ICES, 58
International Decade of Ocean Exploration, 102, 207n101
International Geophysical Year (IGY), 115
International Indian Ocean Expedition, 115, 206n85
International Ocean Institute, 148
International Peace Institute, 67. *See also* Albert I of Monaco
Iron seeding, 158, 162–163
Iselin, Columbus, 99

Jackson, Henry M., 121–122
Japan: and American expansion in the Pacific, 81; Japanese marine investigations, 83; Japanese visitors at the Golden Gate exposition, 91
Jasanoff, Sheila, 8
Jeffreys, John Gwyn, 35
John Pennekamp Coral Reef State Park, 147
Johnson, Lyndon B., 94, 111, 122
Joubin, Louis, 68, 198n108

Kennedy, John F., 93, 99, 131, 219n136
Kent, Rockwell, 1

Kipling, Rudyard, 20
Kishinouye, Kamakichi, 82–83
Knappen, Theodore M., 77
Knudsen, Martin, 194nn69–70
Kofoid, Charles, 56, 198n117
Kohler, Robert, 7

Lacaze-Duthiers, Henri, 46–47, 50–57, 166, 189n10, 190nn16–17, 191n22, 191n32, 191n34
Laloë, Anne-Flore. *See* Shipped perspective
Langston, Nancy, 171
Lankester, Edwin Ray, 42–43, 188n143, 192n46
Lanphier, Edward H., 108
Latour, Bruno, 164
Le Roux, Jeanne, 196n94
Lewis and Clark expedition, 80, 127
Lightning (HMSS), 5, 31–32
Limbaugh, Conrad, 105
Lindbergh, Jon, 129
Link, Edwin, 108, 129
Liquid-breathing experiments, 108, 211n38
Lyman, Theodore, 33

Magnuson, Warren G., 95, 101, 119, 121–122
Mahnken, Conrad, 124
Maienschein, Jane, 189n9
Makai Test Range. *See* Pryor, Tap
Manganese nodules, 153
Man-in-the-sea, 101, 108, 110, 117, 119. *See also* Saturation diving
Man-in-the-Sound, 130
Mann, Thomas, 139, 145, 149
Mare Liberum. *See* Grotius, Hugo
Margerie, Emmanuel de, 70
Marine algae, 29, 51, 106, 158
Marine Biological Association, 43
Marine Biological Laboratory (MBL), 92–93, 175n5, 189n9, 205n62

Marine environment: changing perceptions of the, 138–139; infrastructure in the, 156
Marine geology, 97, 120, 124, 131, 208n8
Marine microbiology, 189n4
Marine protected areas, 136, 165, 169
Marine sciences: act, 100; field and laboratory boundary, 5, 11, 49, 152, 159; international cooperation in, 8–10, 49, 64, 100, 115, 150, 160; paradox of French, 57–60
Marine stations, 9; and university instruction, 49, 53–54; Arcachon marine station, 37, 50; Banyuls-sur-Mer marine station, 50, 53–54, 56, 68; Concarneau marine station, 49, 68; Lowestoft marine station, 43; marine station movement, 9, 47–50; Millport marine station, 43; Misaki marine station, 84, 92; Plymouth marine station, 43, 157, 188n143; Roscoff marine station, 50–57, 68–70, 193n53, 198n117; Sébastopol marine station, 50; Tatihou marine station, 54, 68, 192n38; Wimereux marine station, 54, 57. *See also* Marine Biological Laboratory (MBL); Naples Biological Laboratory; Scripps Institution of Oceanography.
Maritime fiction, 25–26
Marsigli, Luigi-Ferdinando, 15–17, 179n14
Martin, John, 158
Matsen, Brad, 107
Maurice Ewing (RV), 159
Maury, Matthew Fontaine, 13, 24–25, 80–81, 139, 166, 181n55; international maritime conference, 24–25, 166
McConnell, Anita, 175n5, 179n14
McCray, Patrick, 8
McEwen, George, 82–84
McNeill, J. R., 170
Melville, Herman, 1, 25, 135

Menard, Henry William, 1, 150
Merrie, Andrew. *See* Radical Ocean Futures
Merrill, Spencer, 128
Merz, Alfred, 76. *See also Meteor* expedition
Meteor expedition, 5, 76, 201nn10–11
Meteorology, 7, 13, 58, 94, 99, 197n96, 197n103
Mills, Eric, 4, 83, 181n39
Milne-Edwards, Alphonse, 31, 62, 189n10, 195n84, 196n87
Moore, Matt, 169
Morcos, Selim, 57
Morse, Robert W., 99
Morus, Iwan Rhys, 2
Munk, Walter, 98
Murray, John, 32, 35, 68, 76, 188n148
Musée Océanographique, 9, 65–69, 72, 106, 197n103, 198n110, 210n26
Museum of Natural History (Jardin des Plantes), 50, 53, 62–63, 68
Mutsu incident, 138

Nansen, Fridtjof, 56, 68, 193n59, 194n69, 197n97
Naples Biological Laboratory, 47, 50, 53–57, 189n9, 191n34; choice of location, 55–56; developmental biology research, 57, 193n62
Nares, George, 31
NASA, 94, 106–107, 123, 133, 159, 219n142; moon landing, 109
National Petroleum Council, 113–114
Naval Undersea Warfare Center, 117, 229n4
Newfoundland, 136–137, 141, 212n52; Grand Banks, 40, 63, 137, 141–142, 212n51
Nierenberg, William A., 150
NOAA, 111, 118–120, 133, 159. *See also* Oceanlab
Norman, Alfred Merle, 34
North, Wheeler, 105
North Sea Convention, 58

Northwest Passage, 154–155
Norway, 22, 42, 58, 64, 176n14, 194n67, n69, n70

Obama, Barack, 169
Ocean: colonization, 11, 102, 109–111; travel, 19–20; vengeful, 161
Ocean Cleanup Project, 163
Oceanic turn, 6, 167
Oceanlab, 133, 219n143
Oceanographic Commission of Washington, 121–122, 125, 127
Oceanography: history of, 3–8; marine biology/zoology, 3–4, 9, 22, 30, 43, 49–50, 54, 58–59, 181n39; physical of oceanography in France, 54, 58, 192n42, 199n122; physical oceanography and warfare, 103; transformation into global systems science, 150–151
Oceans Observatory Initiative, 159
Office of Naval Research, 98, 103, 126
Oil spills, 147
Okeanus Explorer (RV), 159
Our New Age. *See* Spilhaus, Athelstan
Outer Continental Shelf Lands Act, 112

Pacifica (statue), 87–88
Pacific House, 87, 89, 90, 97. *See also* Youtz, Philip N.
Pacific Remote Islands National Monument, 169
Pacific Science Center, 93, 97
Pacific Science Congresses: Fifth Pacific Science Congress, 78–79; First Pacific Science Congress, 79, 85; Seventh Pacific Science Congress, 91; Sixth Pacific Science Congress, 77–79, 91
Panama Canal, 75, 85–86, 141–142, 155
Panama-Pacific International Exposition, 85–87; "submarines" fairground attraction, 86–87
Papua New Guinea, 153, 157
Pardo, Arvid, 147–148, 167, 223n48

Paul, Harry, 51, 191n30
Pauly, Daniel, 136
Payne, Roger, 147
Perrier, Edmond, 44, 54, 68
Pettersson, Otto, 59–67, 194
Phillips, Denise, 2
Piccard, Jacques, 211n42
Pirate radio stations, 113, 213n62
Plankton, 69, 83, 157–158, 162, 165, 188n148, 226n90
Plastic pollution, 136, 163, 170, 220n5, 229n128
Porcupine (HMSS), 5, 31–32
Portier, Paul, 68
Princesse Alice, 64, 66
Project Argus, 96
Project Artemis, 218n135
Project Blue Water, 118
Project Bottom-Fix, 111
Project Horizon, 214n84
Project Iceberg, 134
Project Natick, 225n75
Project Sea Use, 11, 101–102, 112, 117, 121–128, 134, 217n111; chain highway, 131; diver selection criteria, 129; legality of territorial claim, 117; seamount instrumentation, 130
Pryor, Tap, 217n112
Putnam, Sarah Gooll, 27–28

Radical Ocean Futures, 154. *See also* Stålenhag, Simon
Ray, Dixie Lee, 97, 138
Ray, Louis M. *See* Grand Capri Republic
Rehbock, Philip, 23
Remote sensing, 152, 158–160, 230n8; Argo floats, 159; ocean observatories, 158–159
Resvoll-Holmsen, Hanna, 196n94
Revelle, Roger, 74, 98, 103, 105, 156, 207n103, 226n85
Richard, Jules, 65, 68, 72, 197n99
Riker, Carroll Livingston, 139–140, 149, 160; International Police of the Seas, 142; Mississippi flood control, 141, 143

Ritter, William Emerson, 84
Robb, Walter L., 108. *See also* Liquid-breathing experiments
Roosevelt, Franklin Delano, 90
Rouch, Jules, 3
Roux, Wilhelm, 57
Rozwadowski, Helen, 5–6, 80, 101
Rusk, Dean, 94
Rydell, Robert, 86

Sands, Walter, 120
Sars, Michael, 21, 31, 33
Satellite oceanography, 152, 158–160. *See also* Seasat A
Saturation diving, 102, 108, 118, 132, 211n33, 212n46. *See also* Man-in-the-sea
Sauerwein, Charles, 70, 199n122. *See also* Thoulet, Julien
Scandinavian fisheries, 58, 176n14
Schmidt, Oscar, 33
Schuster, Arthur, 46
Scientific internationalism, 10–11; in a Pacific World, 78, 82–86
Scripps Institution of Oceanography, 11–12, 82–84, 90, 98, 103, 105, 109, 126, 156, 201n20, 202n30
Scuba, 11, 101–106, 131, 167; diving clubs, 104; hyperbaric chamber experiments, 128; tinkering and diving innovation, 124
Seafloor mining, 110, 114, 135, 153, 155, 212n45, 219n139, 224n67
Sealab habitat program, 110, 131, 212n46. *See also* Sheats, Robert
Seamounts, 110, 118, 122; Cobb Seamount discovery, 119–120; Gregg Seamount, 212n51. *See also* Project Sea Use
Seasat A, 159. *See also* Satellite oceanography
Seasteading Institute, 154, 224n70
Seattle World's Fair, 10, 93–95
Sensor buoys. *See* Remote sensing

Sheats, Robert, 131, 218n133. *See also* Sealab habitat program
Shipped perspective, 150
Shor, Elizabeth Noble, 12
SkyTruth, 160
Sociotechnical imaginaries, 8
Sounding/depth sounding, 15–18, 21, 24, 32, 70, 131, 180n35
South China Sea, 134, 136
Soviet saturation diving, 118; Cold War submarine strategy, 132
Spilhaus, Athelstan, 94–97, 205n71, 208n5. *See also* Bathythermograph
Stålenhag, Simon, 153–154. *See also* Radical Ocean Futures
Stanley Jr., Emory Day, 126, 217n117
Steamships, 19; passenger steamships, 20, 30; steam-powered oceanographic vessels, 43, 65, 75, 83; steam trawlers, 39–41, 71, 186n128, 187n133
Stettinius, Edward, 91
Stevenson, Robert Louis, 25
Stommel, Henry, 140, 151
Stoye, John, 16
Sturken, Marita, 104
Submarine telegraphy, 19, 20, 23, 26, 31, 51, 70, 75
Submarine warfare, 73, 94, 98, 102
Submersibles/submarines: Cousteau's diving-saucer, 97; *Jiaolong*, 134; nuclear submarines, 98, 102, 151; *Sea Otter*, 130; *Severyanka*, 214n83
Svalbard, 64
Sverdrup, Harald, 90

Talisman and *Travailleur*, 5, 47–49, 62–63, 68
Taylor, Walter P., 86
Teas, Howard, 105
Technology, transformation of the seas by, 19–21
Tektite habitat program, 110, 123–125
Tharp, Marie, 198n118

Thiel, Peter, 154. *See also* Seasteading Institute
Thomas, Douglas, 104
Thomson, Charles Wyville, 19, 30–35, 75, 185n101, 185n107
Thoulet, Julien, 40, 59, 68, 70, 194n70, 195n77, 199n122
Three-nautical-mile territorial limit, 112
Thresher and *Scorpion* sinking, 122
Tides, 16–17
Titly, David, 154
Truman, Harry S., 112. *See also* Outer Continental Shelf Lands Act
Trump, Donald, 169
Turbot war, 136–137
Turley, Carol, 157

United Nations, 91, 113, 136, 145, 168–169, 213n73; UNESCO, 115, 170; Convention on the Law of the Sea (UNCLOS), 147, 223n48
US Coast Survey, 24, 32, 182
US Exploring Expedition, 24, 80. *See also* Wilkes, Charles
US Fish Commission, 25, 40, 47. *See also* Baird, Spencer F.
Ussner, Alexander, 27

Valéry, Paul, 1
VanDerwalker, John, 124
Van Oppen, Madeleine, 164
Vaughan, Thomas Wayland, 82, 202, 208n8
Verne, Jules, 13, 25, 109, 181–182n55
Virginia Mason Medical Center, 128–129. *See also* Campbell, Spence

Weather control, 94, 141, 215n91
Weber, Eugen, 51
Whales: adaptation to ship noise, 155; electrocardiogram measurement;

whales exhibited, 27–29, 38; whaling, 24, 77, 160. *See also* Animal tracking; White, Paul Dudley
White, Paul Dudley, 160
White, Roland, 129
Wilhelm II, Kaiser, 71, 77, 198, 199n124
Wilkes, Charles, 24, 80. *See also* US Exploring Expedition
Wilson, Woodrow, 71, 75, 212–213n58
Woods Hole Oceanographic Institution, 11, 94, 99, 208n5
Work, Hubert, 74
World Federalist Movement, 145
World War I, 10, 73, 76, 99, 112, 166, 194n72, 212–213n58
World War II, 10, 79, 90–97, 106, 138, 145, 167, 178n30, 205n61; attack on Pearl Harbor, 90; Treasure Island naval base, 90

Yachting, 14, 20, 25
Yamasaki, Minoru. *See* Federal Science Exhibit
Youtz, Philip N., 79, 90. *See also* Pacific House